P9-ASH-966

THE FACTS ON FILE
PHYSICS
HANDBOOK

THE FACTS ON FILE
PHYSICS
HANDBOOK

THE DIAGRAM GROUP

Facts On File, Inc.

The Facts On File Physics Handbook

Diagram Visual Information Ltd

Editorial director Moira Johnston
Editors Nancy Bailey, Jean Brady, Paul Copperwaite, Eve Daintith, Bridget Giles, Jane Johnson, Reet Nelis, Jamie Stokes
Design Richard Hummerstone, Edward Kinsey,
Design production Carole Dease, Oscar Lobban, Lee Lawrence
Artists Susan Kinsey, Lee Lawrence, Kathleen McDougal

Research Peter Dease, Catherine & Neil McKenna,
Contributors Michael Allaby, Martyn Bramwell, John Daintith, Trevor Day, John Haywood, Jim Henderson, David Lambert, Catherine Riches, Dr Robert Youngson

Indexer Christine Ivamy

Facts On File, Inc.
132 West 31st Street
New York NY 10001

Library of Congress Cataloging-in-Publication Data

The Facts on File physics handbook / The Diagram Group.
p. cm.
Includes index
ISBN 0-8160-4082-6
I. Physics—Handbooks, manuals, etc. I. Diagram Group.

QE5 .F32 2000
550—dc21

99-048564

Facts On File books are available at special discounts when purchased in bulk quantities for businesses, associations, institutions, or sales promotions. Please call our Special Sales Department in New York at 212/967-8800 or 800/322-8755.

You can find Facts On File on the World Wide Web at http://www.factsonfile.com

Cover design by Cathy Rincon

Printed in the United States of America

MP DIAG 10 9 8 7 6 5 4 3

This book is printed on acid-free paper.

INTRODUCTION

An understanding of science is the basis of all technological advances. Our domestic lives, possessions, cities, and industries have only been developed through scientific research into the principles that underpin the physical world. But obtaining a full view of any branch of science may be difficult without resorting to a range of books. Dictionaries of terms, encyclopedias of facts, biographical dictionaries, chronologies of scientific events – all these collections of facts usually encompass a range of science subjects. THE FACTS ON FILE HANDBOOK LIBRARY covers four major scientific areas – CHEMISTRY, PHYSICS, EARTH SCIENCE (including astronomy), and BIOLOGY.

THE FACTS ON FILE PHYSICS HANDBOOK contains four sections – a glossary of terms, biographies of personalities, a chronology of events, essential charts and tables, and finally an index.

GLOSSARY
The specialized words used in any science subject mean that students need a glossary in order to understand the processes involved. THE FACTS ON FILE PHYSICS HANDBOOK glossary contains more than 1,400 entries, often accompanied by labeled diagrams to help clarify the meanings.

BIOGRAPHIES
The giants of science – Darwin, Galileo, Einstein, Marie Curie – are widely known, but hundreds of other dedicated scientists have also contributed to the advancement of scientific knowledge. THE FACTS ON FILE PHYSICS HANDBOOK contains biographies of more than 300 people, many of whose achievements may have gone unnoticed but whose discoveries have pushed forward the world's understanding of physics.

CHRONOLOGY
Scientific discoveries often have no immediate impact. Nevertheless, their effects can influence our lives more than wars, political changes, and world rulers. THE FACTS ON FILE PHYSICS HANDBOOK covers nearly 9,000 years of events in the history of discoveries in physics.

CHARTS & TABLES
Basic information on any subject can be hard to find, and books tend to be descriptive. THE FACTS ON FILE PHYSICS HANDBOOK puts together key charts and tables for easy reference.
Scientific discoveries mean that any compilation of facts can never be comprehensive. Nevertheless, this assembly of current information on the subject offers an important resource for today's students.

In past centuries scientists were curious about a wide range of sciences. Today, with disciplines so independent, students of one subject rarely learn much about others. THE FACTS ON FILE HANDBOOKS enable students to compare knowledge in chemistry, physics, earth science, and biology, to put each subject in context, and to underline the close connections between all the sciences.

CONTENTS

SECTION ONE GLOSSARY

A(r) The symbol for relative atomic mass.

ab- An abbreviation for "absolute" and a prefix applied to practical electrical units, such as volt (abvolt), ampere (abampere), or coulomb (abcoulomb), to convert them to units in the absolute electromagnetic system of units.

aberration A defect in the image produced by a lens or mirror caused when rays do not converge accurately to the focus. It is technically difficult to produce a lens or mirror without aberration. *See also* chromatic aberration, spherical aberration.

absolute code Computer program code in a form that can be used directly by the central processing unit. More commonly known as machine code.

absolute zero The lowest possible temperature. Zero on the Kelvin scale or –273°C (–460°F). Absolute zero is unattainable, but temperatures within one millionth of a degree have been reached. At temperatures approaching absolute zero matter exhibits some unexpected properties. Electrical conductors become superconducting and some fluids lose their viscosity (superfluidity). In the Kelvin scale, which has no negative figures, the symbol for degrees in not used after the number. So the freezing point of water is 273K, not 273°K.

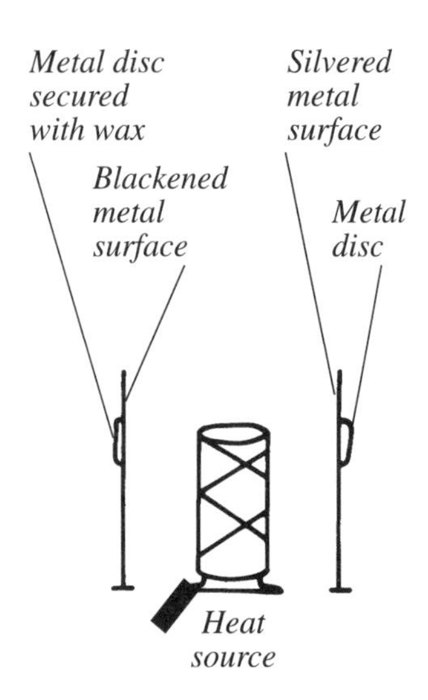

Absorption factor

absorbed dose (of ionizing radiation) The energy absorbed in body tissue by ionizing radiation; unit of measurement, the gray (joule per gram).

absorption (1) The assimilation of radiation into a body with its partial or complete conversion into another form of energy (such as heat, light, sound, etc.).
(2) The way in which certain wavelengths in a beam of radiation are removed when passing through a substance. The resulting spectrum of radiation can be used in spectroscopy to analyze the substance.
(3) The process by which one material, e.g., water, is taken up by another, e.g., a sponge.

absorption factor Energy absorbed by a body depends on the nature of its surface. Shiny, light-colored surfaces absorb less than dark, matte surfaces.

absorption spectrum The spectrum that forms when radiation with a range of wavelengths passes through a sample. Some wavelengths are absorbed by the sample, leaving gaps in the spectrum.

AC Abbreviation for alternating current.

acceleration A measure of how the velocity of a body changes with time. It occurs if there is a change in the body's speed or direction of motion; a force is required to effect such a change. A falling body whose motion is solely governed by the downward force of gravity is said to be in free fall. The body moves with constant acceleration. In SI units, acceleration is expressed in meters per second per second. It can be calculated by subtracting the final velocity from the initial velocity and dividing the result by the time. Acceleration is a vector quantity. If velocity is plotted against time on a graph, the acceleration at any moment is given by the slope of the graph at that moment.

acceptors Atoms of an element such as boron added to a silicon lattice to affect the conducting properties of semiconductors. The acceptor atoms create holes in the valence band to carry current; this is a p-type (positive carrier) semiconductor.

access time The time taken to retrieve a small item of data from any form of storage in, or connected to, a computer. Access time from RAM may be a very small fraction of a second; from a large magnetic tape store it may be several minutes. Floppy disk drives have longer access times than hard disks, and most CD-ROM drives also have longer access times than hard disks.

accommodation The adjustment of the focus of the eye so that, whatever the range of gaze, a sharp image is formed on the retina. Accommodation is achieved by the internal crystalline lens, which in young people is naturally so elastic that it would tend to approximate to a sphere but for the pull all around its equator by delicate ligaments that suspend it from a circular muscle. When this muscle contracts, the circle becomes smaller and the pull on the lens is less, thus making it a more powerful converger of light and allowing focusing of the image from near objects. When the ciliary muscle relaxes, more distant objects can be focused. The stimulus for accommodation is the degree of convergence of the rays of light, which can be detected by the variations in depth to which the rays penetrate the light-sensitive cells of the retina.

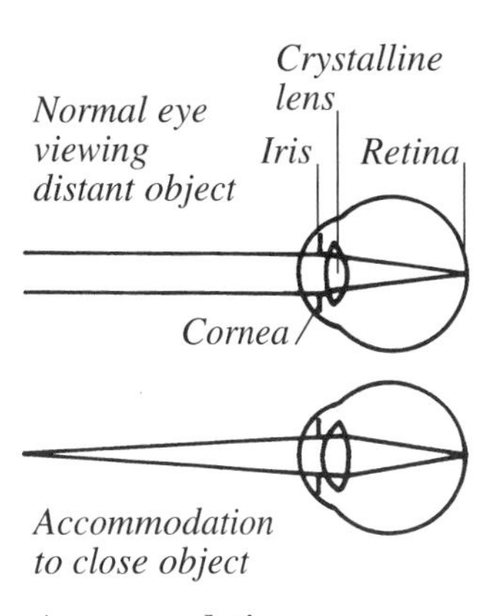

Accommodation

accumulator (1) An electric battery in which the passage of an electric current from an external source brings about a reversible chemical change by which energy is stored. When the electrodes are joined, a current flows and the previous chemical state is restored.
(2) In computing, an accumulator is a small memory or buffer that briefly holds the results of successive operations by the arithmetic and logic unit of the central processing unit.

AC electric motor An electric motor that operates from an alternating electric current supply.

achromatic lens A lens that is free from chromatic aberration. An achromatic lens is constructed from two types of glass, each having a different refractive index, so that the dispersions cancel each other. This form of double lens has the effect of reducing the dispersion of light of different wavelengths, producing a sharper focus with greatly reduced color fringes around the edges of the image.

acoustic coupler An early form of modem that can be temporarily attached to a normal telephone. The digital signal to be transmitted is converted into a two-tone sound that is emitted by a small loudspeaker pressed to the telephone microphone. The received signal is picked up by a microphone held close to the telephone earpiece. The system allows modem communication from any location where there is a telephone, but the speed of transmission is low. Acoustic couplers are now seldom used.

actinic radiation Radiation, especially light and ultraviolet radiation, that can cause a chemical change, such as the latent image on a photographic emulsion or the breakdown of biological or other molecules. The Sun is a major source of actinic electromagnetic radiation.

actinium series A series of radioactive elements each of which derives from the radioactive decay of its predecessor and ends in a stable atom. The actinium series starts with uranium-235 and ends in an isotope of lead. Two other such series occur naturally, one starting with uranium-238 (uranium series) and the other with thorium-232 (thorium series), both also ending with an isotope of lead.

action at a distance Forces such as gravity and magnetism can affect objects not in direct contact with the force. The space through which the forces operate is called a field. For example, a gravitational field is a region in which there are gravitational forces.

activation energy The energy barrier to be overcome in order for a reaction to occur. Many chemical reactions require heat energy to be applied to reactants in order to initiate a reaction.

active device A component in an electronic circuit that consumes power to effect a change, such as amplification or current switching. Transistors are active devices; resistors, capacitors, and inductors are passive devices that do not consume power.

activity (of a radioactive source) The number of ionizing particles emitted per second from a radioactive source, measured in becquerels.

activity series A listing of elements where the position of an element shows how reactive it is. Its position is governed by the ease with which the element releases electrons.

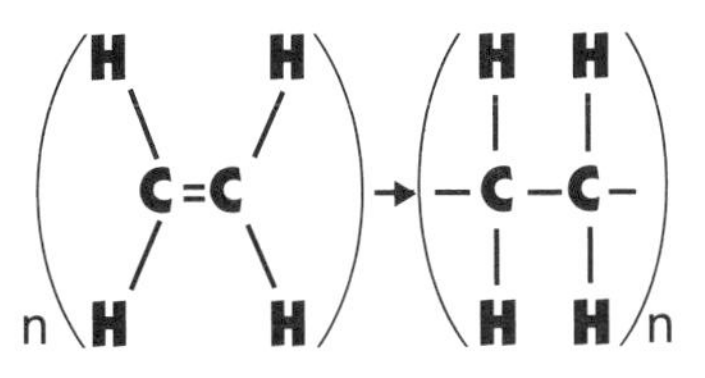

Addition polymerization

addition reaction An addition reaction occurs when at least two molecules react together to form a single molecule (e.g., the polymerization of alkenes).

address A known location in a computer memory at which a single item of data, usually a byte, can be found. The address is simply a number that identifies the location. A computer instruction might be to take the number held in a particular address, multiply it by the number in another address, and put the answer in a third address. An address bus is a set of wires along which signals travel to reach addresses. Data can then be taken from these addresses by way of a data bus.

adiabatic change A change that occurs without thermal transfer to a system. When a change of volume is made very quickly, without allowing any transfer of thermal energy, the change in volume is an adiabatic change. Adiabatic work is work done on a system that does not gain or lose heat to the surroundings.

admittance The reciprocal of impedance.

adsorption The process by which molecules of gases, liquids, or finely divided solids become attached to the surface of another substance to form a very thin layer, often only one molecule thick. Adsorption involves interaction between molecules at a surface. Powdered activated charcoal is a highly effective adsorbent and will remove color from liquids, poisons from liquids in the intestines, and poisonous gases from air passed through it. In absorption, the absorbed substance is taken up by the absorbing substance and distributed throughout it, in the manner of blotting paper drawing up water by capillary action.

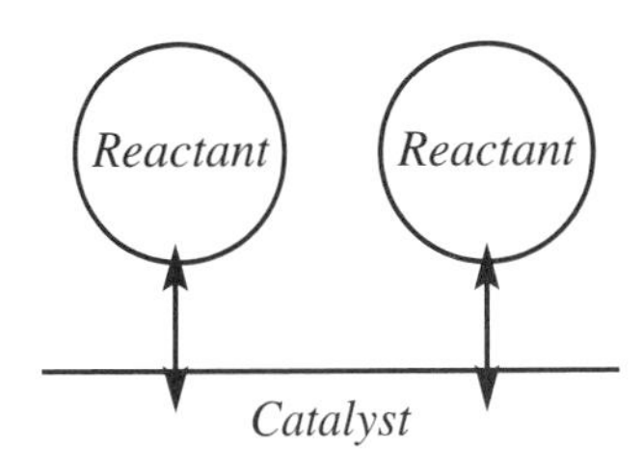

Adsorption

aeration The purification of a substance by exposure to the mechanical or chemical action of air.

aerial The part of a radio or television system that transmits or receives radio waves. *See* antenna.

aerodynamics The study of the dynamic motion of gases, particularly the branch of science studying the motion and control of bodies in air.

affinity The attraction of two substances for each other, leading to the substances combining together.

algol A high-level computer programming language designed primarily for producing problem-solving programs for mathematical and scientific use. The name is an abbreviation of the phrase "algorithmic language." Algol passed though several generations, introduced a number of important new concepts, and was highly influential in the development of programming languages.

algorithm A sequence of instructions to be followed with the intention of finding a solution to a problem. Each step must specify precisely what action is to be taken and, although there may be many alternative routes through the algorithm, there is only one starting point and one end point. Various nodes occur at which decisions must be made and these are decided by questions that can be answered "yes," or "no." The direction taken at these nodes is determined by the answer. A computer program commonly involves algorithms, and the preliminary studies for a program are often expressed in a particular set of symbols known as a flow chart. This is also an algorithm.

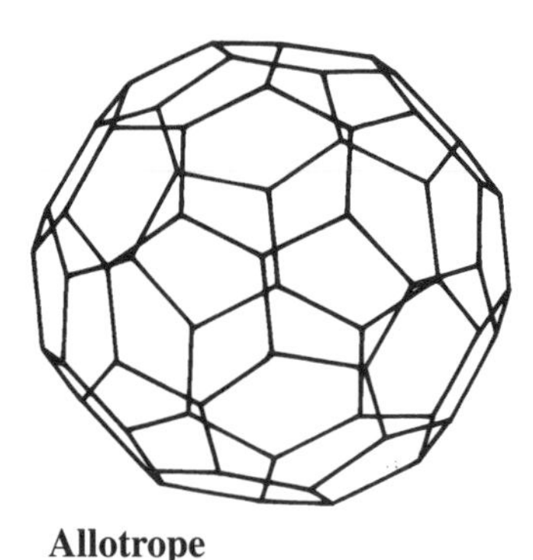

Allotrope (buckminsterfullerene)

allotrope An element that can exist in different physical forms while in the same state. Carbon can occur as two common allotropes, diamond and graphite. (A third, buckminsterfullerene, has been discovered recently.) The physical properties of these allotropes are very different.

alloy A metallic material made of two or more metals or of a metal and nonmetal. By mixing metals in certain proportions, alloys with specific properties can be made.

alphanumeric characters The 26 letters of the alphabet and the decimal digits 0 to 9. Each of these, and many other characters, are represented in computers by a code such as the ASCII code.

alpha particle (α particle) A particle released during radioactive decay. It consists of two neutrons and two protons and is the equivalent of the nucleus of a helium atom. Energy is released by this change, most of it accounted for by the kinetic energy of the alpha particle, which moves away at high speed but which rapidly loses energy by collision and ionization of other atoms and molecules and is easily stopped by a piece of paper. Their range in air is only a few centimeters and shielding against them is easy. Alpha rays are streams of fast-moving alpha particles.

alternating current (AC) A continuous electric current that varies in strength, regularly reversing its direction.

amines A group of organic compounds containing the amino functional group $-NH_2$.

amino acid A group of organic compounds containing both the carboxyl group (–COOH) and the amino group ($-NH_2$).

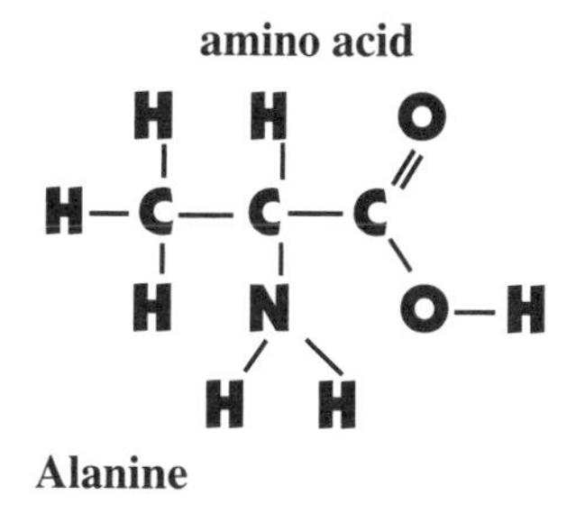

Alanine

amorphous Lacking form, shape, or crystal structure; amorphous substances have no fixed melting point.

ampere SI unit of electrical current, equal to one coulomb per second.

amphoteric Exhibiting properties of both an acid and a base. An amphoteric compound reacts with both acids and bases to form salts.

amplifier An electronic device for increasing the strength of a varying electrical signal, ideally with minimal alteration to its characteristics (minimal distortion). Amplifiers use low-power transistors to amplify voltage and power transistors to amplify current. Many amplifiers now consist of integrated circuits except for the power output stages. Negative feedback is used to reduce distortion and increase the frequency range over which the amplifier will work well.

amplitude The size of the maximum displacement from the equilibrium position of an oscillation or wave. For an alternating current, the amplitude is the peak value of the current. For a pendulum, the amplitude is half the length of the swing.

amplitude modulation Used in radio transmissions in the long, medium, and short wavebands. A second wave motion is added to a carrier wave to carry the signal to a receiver. The characteristics of this audio-frequency signal vary the amplitude of the carrier wave.

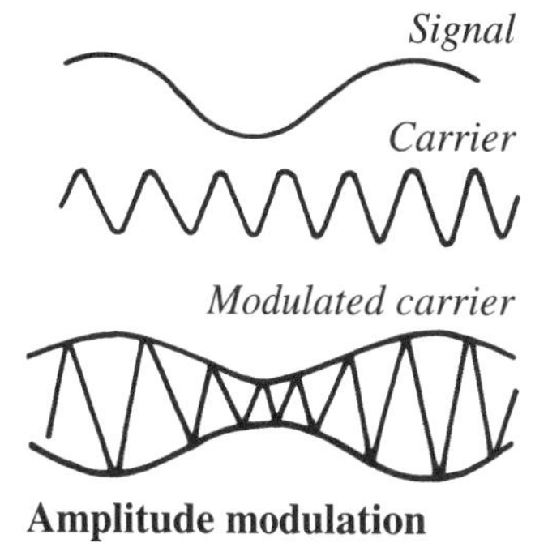

Amplitude modulation

analog In physical science, a continuous representation, of any kind, of a varying quantity. Thus, the movement of the needle of an electric meter over the scale as the current varies is an analog representation of changing voltage. *Compare* digital.

analog to digital converter An electronic device that converts a varying signal, especially a voltage, to a series of numbers. This is done by sampling the signal repeatedly at short intervals, usually many times a second, and expressing each momentary magnitude as a number that can be represented in the form of binary digits. Since almost all information processing is now done by digital computers and almost all information derived from natural sources is analog, the importance of analog to digital conversion can be appreciated.

anechoic chamber An irregularly shaped room in which the walls are covered with small cones or wedges of sound-absorbent material so as to avoid the formation of stationary waves and hence echoes and resonances. Anechoic chambers are acoustically "dead" and are used to test various instruments, such as microphones, and to conduct research into noise and other acoustical phenomena.

anemometer An instrument for measuring windspeed.

angle A measure of the space between two straight lines diverging from a common point of contact.

angle of declination The angle made by a compass needle with the direction of the geographic North Pole, giving a measure of the difference between the geographic and magnetic poles.

angle of deviation The angle between the incident ray and the refracted ray when a light ray passes from one medium to another.

angle of dip The angle made by a suspended magnetic needle to the horizon.

angstrom A very small unit of length equal to one tenth of a nanometer, or one hundred millionth of a centimeter. The angstrom is used to represent the wavelength of radiation at the short wavelength (high frequency) end of the electromagnetic spectrum. It was named after the Swedish physicist Anders Ångstrom. The angstrom has now been largely replaced by the nanometer.

angular acceleration A measure of how the angular velocity of a spinning body changes with time. The rate of change of angular velocity.

angular magnification The ratio of the angle formed at the eye by the final image to that formed at the eye by the object. This is also known as the magnifying power of the system. Linear magnification, on the other hand, is the ratio of the height of the image to that of the object.

angular momentum The angular velocity of an object multiplied by its moment of inertia (i.e., in the case of a simple rotating wheel, multiplied by its mass multiplied by the square of the distance from its axis of rotation). It is the product of rotational inertia and angular velocity.

Angular momentum

angular velocity The rate at which a rotating body moves through an angle about an axis. Speed of motion in a circle, or, more precisely, the rate of change of angular displacement with time.

anhydrous Containing no water, a term applied to salts without water of crystalization.

anion An ion having negative charge.

anode The electrode carrying the positive charge in a solution undergoing electrolysis. A positive electrode toward which negative particles, such as electrons or negative ions, are attracted. Negative ions are called anions. In an electric cell or battery, the anode is the electrode that attracts electrons to itself from the external circuit.

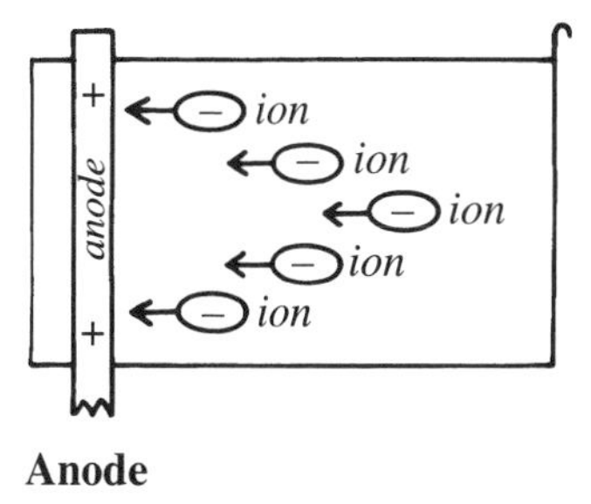

Anode

ANSI Abbreviation for American National Standards Institute, a body that lays down various standards for computers, computer connections, connecting pin positions, disk and tape drives, software, and so on. Many ANSI standards are observed worldwide, and many personal computers have in their operating system directory an ANSI.SYS file that can be read by the machine at the time of switching on.

antenna or **aerial** An electrical conductor, taking a variety of forms, from which radio signals are transmitted or by which they are received. Antennas may consist of long wires suspended high above the ground, dipoles with twin arms insulated from each other, or short ferrite rods incorporated in small transistor receivers. Often their dimensions are calculated to resonate with the principal wavelength or waveband of interest. They may have reflectors behind them and directors in front of them and may have to be accurately aligned. Antennas for microwaves often consist of a parabolic reflector with the actual antenna set at the focus of the parabola. The gain of an aerial is the degree to which its performance matches that of an approved standard. Transmitting antennas correspond dimensionally to effective receiving antennas but have to handle power and may have thicker conductors.

antimatter Each subatomic particle has its antiparticle, its properties being equal and opposite to those of the particle. Antiparticles make up antimatter. Examples of antiparticles are antineutrons, antiprotons, antineutrinos, and positrons (the antiparticles of electrons). When an antiparticle meets its corresponding particle, both are annihilated and the corresponding energy released as photons of radiant energy. Some antiparticles have been produced by particle accelerator/collider experiments and some are put to practical use, as in the positron emission tomography (PET) scanner. The theoretical grounds for the existence of antimatter were presented by the English physicist Paul Dirac in 1928 as a prediction of relativistic quantum mechanics.

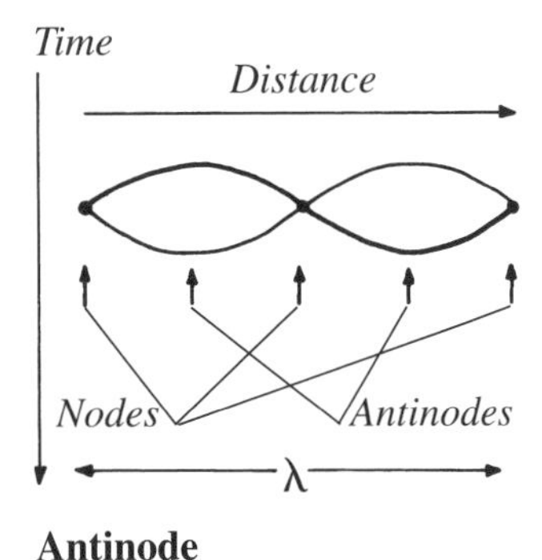

Antinode

antineutrinos *See* antimatter.

antinode In a standing (stationary) wave it is the point of maximum displacement (either positive or negative).

anti-phase Two waves with the same wavelength and frequency are in anti-phase if their phase difference is half a wave.

antiprotons *See* antimatter.

aperture The useful or effective diameter of a lens or curved mirror. In practical usage, especially in photography, the term is taken to mean the ratio of the focal length of the lens or mirror to its effective diameter. The numerical value of this ratio is known as the f-number of a lens or mirror. Thus a camera zoom lens set at a focal length of 50 mm and an effective diameter (aperture) of 25 mm would have an f-number of 2. Set at 100 mm the same aperture would give an f-number of 4 and only half the exposure for a given shutter speed.

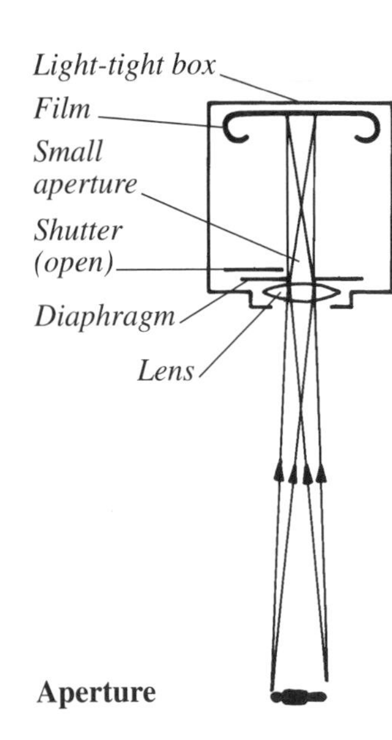

Aperture

applications software Programs that carry out specific functions, such as word processing, spreadsheet operations, database access, computer-assisted design, accountancy, and so on. Applications software is distinguished from computer operating systems.

Archimedes' principle The weight of liquid displaced by a body that is totally or partially immersed in the liquid is equal to the apparent loss of weight of the body.

architectural acoustics The study of the features of buildings and auditoriums that allow music and speech to be heard clearly and comfortably. Good hall acoustics imply the absence of undue echoes or of the emphasis of any particular pitches by resonances. Large, plane, unbroken surfaces reflect sound, and simple dimensions promote resonances. Good acoustics also imply that the reverberation time (the time taken for audible re-echoing to drop to an acceptable level) should be appropriate for the main purpose of the hall. Long reverberation times cause indistinct speech and blurred music; unduly short reverberation times produce a "dead" effect. Bare rooms with hard surfaces increase reverberation times; carpets, soft furnishing, and the presence of an audience reduce reverberation.

archiving The movement of a computer file from a position of immediate access, as on a hard disk, to one of less immediate access, as on a remote tape drive or on a remotely stored disk. Archiving may be performed deliberately by the operator or may occur automatically, after a designated period, as part of a programmed process. Archiving is not the same as backing up.

area A measure of the extent of a surface.

armature (1) Part of an electric machine, such as a bell, that vibrates when a magnetic field is applied.
(2) The moving soft-iron core structure around which wire is coiled in an electric motor.

association A term used in Boolean algebra.

astable circuit An electronic circuit that can be in one of two states, neither of which is permanently stable. Often used as an oscillator. An astable usually consists of two interconnected transistors, one turned on, the other turned off. Each of the pair alternately oscillates between the on and off state and, in so doing, switches the other to change its state. An astable circuit produces a square wave or pulse output and, if synchronized to an accurate frequency by a quartz crystal or other means, can operate as a timing clock for computers or other devices. Quartz watches and clocks use crystal-controlled astables to generate a high-frequency square wave, which is then repeatedly frequency-divided by bistables until a one-second square wave is reached.

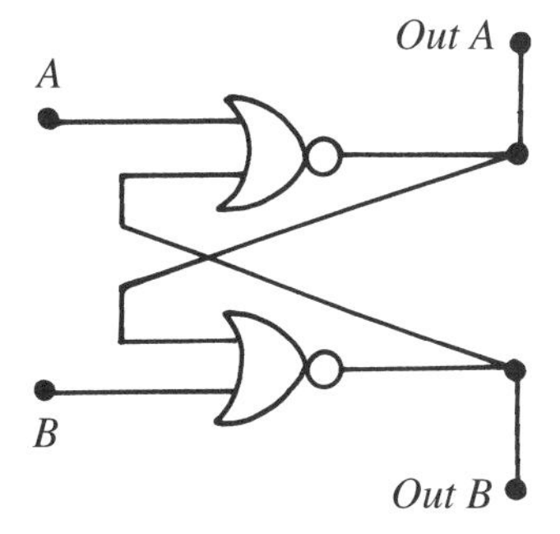

Astable circuit component

astigmatism A property of a lens or mirror system in which a surface is not that of a perfect sphere or parabola, but has a greater degree of curvature in one meridian than in the meridian at right angles to that one. Such a toric surface produces two foci – one for rays in the plane parallel to that of greatest curvature and one for rays in the plane at right angles. Only one of these sets of rays can be focused at one time. The cornea of the eye is commonly astigmatic, usually being most steeply curved from top to bottom and least steeply curved from side to side. Ocular astigmatism is corrected by spectacles having cylindrical lenses set at the appropriate axis so that the steeper corneal curve is matched by the less steep spectacle lens curve.

astrophysics The study of the physics and chemistry of the stars, including their origin, evolution, and structure and the generation and movement of energy in and around them. Astrophysics is also concerned with the relationships and dynamics of star clusters and galaxies. It is also concerned with the study of interstellar dust and molecules, comets, asteroids, and any other extraterrestrial matter.

atmosphere (1) The layer of gases that surround a planet or other heavenly body.
(2) A unit of pressure.

atom The smallest particle of an element that can exhibit that element's properties. An atom has a small massive nucleus of protons and neutrons surrounded by a cloud of electrons (equal in number to the number of protons in the nucleus and unique to the element).

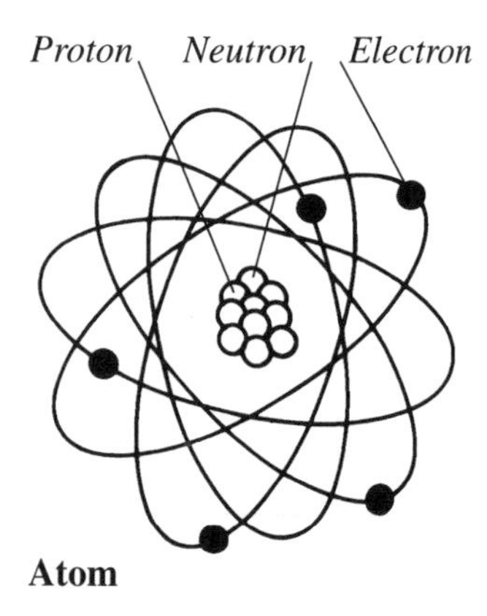

Atom

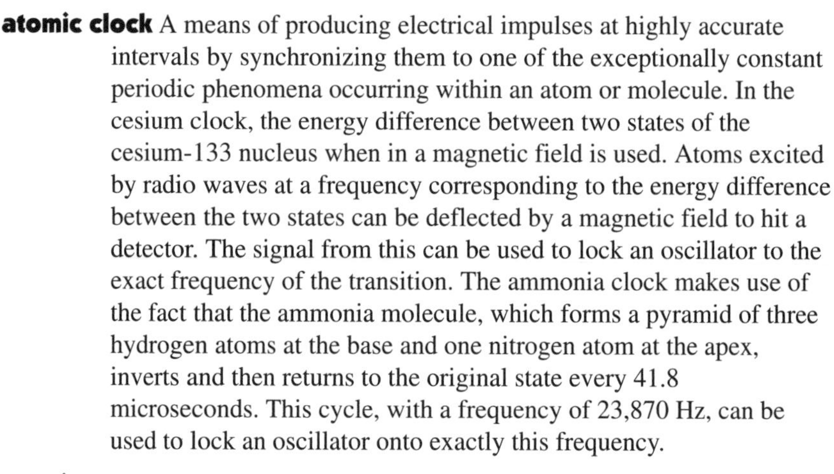

atomic clock A means of producing electrical impulses at highly accurate intervals by synchronizing them to one of the exceptionally constant periodic phenomena occurring within an atom or molecule. In the cesium clock, the energy difference between two states of the cesium-133 nucleus when in a magnetic field is used. Atoms excited by radio waves at a frequency corresponding to the energy difference between the two states can be deflected by a magnetic field to hit a detector. The signal from this can be used to lock an oscillator to the exact frequency of the transition. The ammonia clock makes use of the fact that the ammonia molecule, which forms a pyramid of three hydrogen atoms at the base and one nitrogen atom at the apex, inverts and then returns to the original state every 41.8 microseconds. This cycle, with a frequency of 23,870 Hz, can be used to lock an oscillator onto exactly this frequency.

atomic mass Short for relative atomic mass.

atomic mass unit Defined as 1/12 the mass of one atom of carbon-12 isotope.

atomic number or **proton number** (Z) The number of protons in the nucleus of an atom. If the nucleus is not electrically charged, this is equal to the number of electrons in its shells.

atomic radius A measurement of the size of an atom in nanometers.

atomic spectrum An atomic spectrum shows all electromagnetic emissions from a sample of the element.

avalanche breakdown Uncontrollable breakdown (of insulation) in a semiconductor junction at a very sharply defined voltage level.

Avogadro constant or **number** (L) The number of particles (atoms, molecules, ions) present in a mole of substance. Specifically, it is the number of atoms present in 12 g of the carbon-12 isotope (6.023×10^{23}).

Avogadro's hypothesis or **law** Equal volumes of all gases at the same temperature and pressure contain the same number of molecules.

axiom A proposition generally agreed to be so obviously true as not to require proof; a proposition that appears incapable of proof but is stipulated to be true for purposes of logical argument.

back e.m.f. The e.m.f. (electromotive force) induced in a coil rotating in a magnetic field (as in an electric motor). It opposes the e.m.f. of the supply. The faster the coil rotates, the larger the back e.m.f.

background radiation The reading on a Geiger-Müller tube when it is not near a radioactive source. The reading of about one radiation pulse per second is caused by low-intensity radiation from natural and human-made sources that is constantly present in the environment. This radiation comes from radioactive substances in rocks and soil and from cosmic radiation from outer space. Some of the soil and atmospheric radiation comes from long-lasting fallout from nuclear weapons and from waste gases from nuclear power stations, but the total effect of human-made radiation is very small compared to the natural levels. In most areas, background radiation produces a count of about one per second in a Geiger counter, and this must be taken into account when measuring radiation from a particular source. Microwave background radiation at 2.7K is the residual radiation from the big bang.

backup A copy of important software, especially of data that cannot readily be replaced, made for use in the event of the loss or corruption of the primary source. Some systems automatically make backup copies at intervals that can be specified. It is sound practice to keep backup material in a different building from the primary material.

ballistic galvanometer A type of galvanometer that is used for measuring surges of current.

ballistics The study of the flight dynamics and path taken by projectiles, the propulsive forces and their effect, and the effect of air resistance and gravity.

Balmer series A series of lines between the wavelengths of approximately 365 and 655 nm in the hydrogen emission spectrum.

bandpass filter An electronic or passive device that greatly reduces the amplitude of (attenuates) signals of wavelength outside a prescribed range.

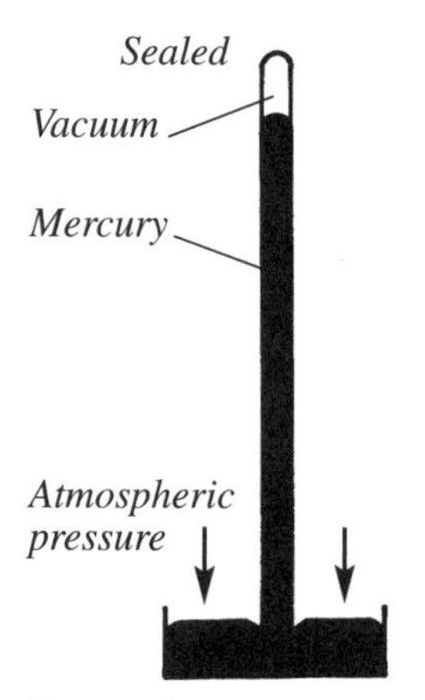

Barometer

bandwidth (1) The range of frequencies (or their reciprocal, wavelengths) over which a system is capable of operating within defined limits of efficiency.
(2) The range of frequencies that a communication channel can accommodate.
(3) The frequency spread of a radio signal of a given nominal frequency (wavelength).

barometer An instrument for measuring atmospheric pressure.

baryon The most massive of the subatomic particles; protons and neutrons are baryons. They consist of three quarks.

base unit A unit of measurement that is defined in terms of physical entities rather than being derived from other units.

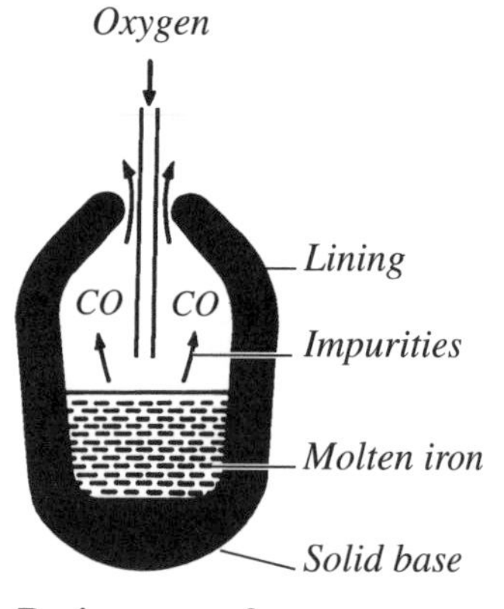

Basic oxygen furnace

basic oxygen furnace (steel) Molten iron passes straight from the blast furnace to a container lined with heat-resistant alkaline (basic) bricks. Scrap steel and limestone are added. Oxygen is blown onto the surface of the molten mixture, and impurities burn off. In order to achieve steels with specific qualities, other elements are added and the resulting steel is cast into long continuous strips.

battery Two or more electrical cells connected in series or parallel.

baud rate The number of binary digits (bits) transmitted per second along a line. More generally, the number of times per second that a data transmission channel changes its state. The term was derived from the name of the French communications engineer J. M. E. Baudot who invented an efficient code that replaced the Morse code for telegraphic signalling.

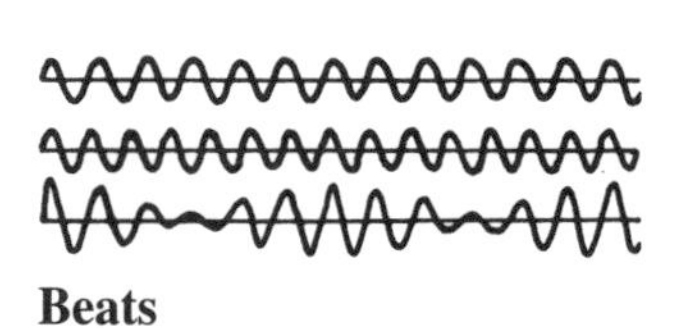
Beats

beat The interference effect between two waves of almost, but not quite, identical frequency. Such waves pass in and out of phase with each other and thus alternately potentiate and reduce each other, giving a resulting wave that varies between maximum intensity, where the waves add together, and minimum intensity, where the waves cancel each other completely. This process gives rise to an additional wave of much lower frequency than either of the beating waves. The beat period is the time between successive beats (or maximum displacements of the wave), and is the inverse of the difference between the two frequencies.

becquerel The SI unit for the activity of a radioactive source. One becquerel is an emission of one particle per second.

Bernoulli effect The effect whereby the pressure exerted by a fluid moving slowly is greater than that caused by a fluid moving quickly. This effect causes lift on an aircraft wing. The air moves faster over the curved upper surface of the wing than over the lower surface of the wing. There is thus a smaller pressure on the upper surface than on the lower leading to a net upward thrust, the lift.

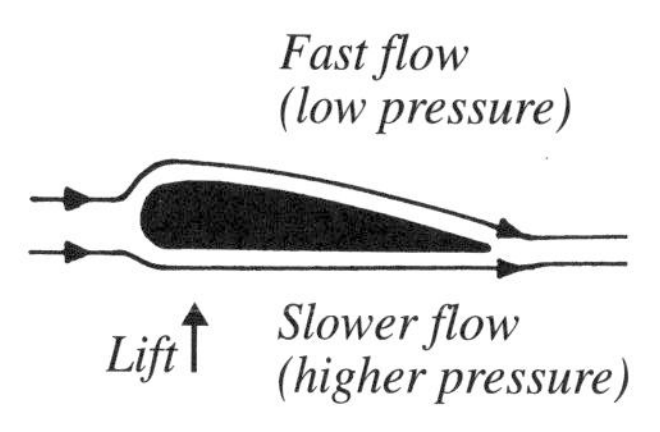

Bernoulli effect

beta particle (β particle) A high-speed electron emitted by the nucleus of certain radioactive elements during beta decay. When a neutron in the nucleus decays to a proton, an electron is emitted and the number of protons increases by one, so the atomic number increases by one. A beta ray is a stream of high-energy electrons. They will produce ions in matter through which they pass and will penetrate a layer of several millimeters of aluminum.

beta testing Evaluation and trial of a new or modified software package in normal working conditions, often by large numbers of experienced users, so that faults can be detected. Beta testing is done prior to the commercial release of the package.

betatron A particle accelerator that produces high-speed electrons for research purposes. The negative particles are accelerated by powerful electromagnetic fields.

big bang The standard model of the universe suggests that between 10 and 20 million years ago the universe began with a gigantic explosion, creating matter, space, and time. The observed movement of galaxies away from each other can thus be seen as the continuation of this explosion.

bimetallic strip A strip of metal made of two different metals joined (riveted or welded) together. When heated, the strip bends because the metals expand by different amounts. Bimetallic strips are used as temperature-control devices.

binary code The system in which numbers are represented in a positional system using the base 2 instead of the more familiar base 10 (decimal system). The only integers used are 0 and 1.

binding energy The energy equivalent to the loss of mass when protons and/or neutrons bind together to form the nucleus of an atom. When these particles bind to form a nucleus, they enter a more stable, lower-energy state. This energy can be released both by splitting (fission) of heavy elements and fusion of light elements.

binoculars An instrument used by both eyes for viewing distant objects. Each

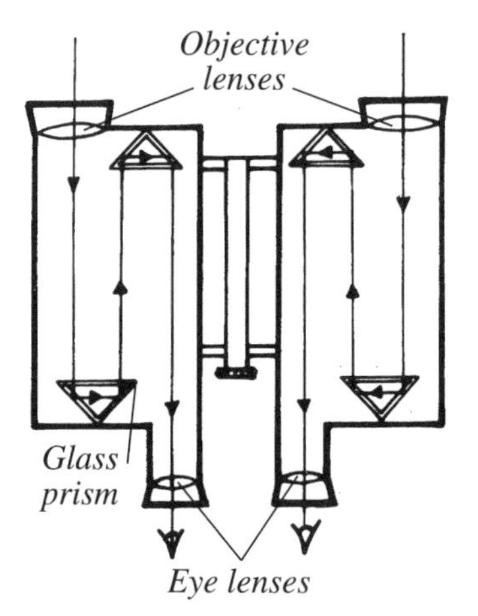

Binoculars

eyepiece is a telescope with a lens at one end by the eye and an objective lens at the other. An arrangement of two right-angled prisms in each tube allows the tube to be short. As each eye has an image, distance perception is obtained.

bionics The study of the relationship between brain function and electronics, or the engineering discipline concerned with the design and production of artificial and functional body parts, organs, limbs, walking devices, and so on.

biophysics The study of those aspects of physics that apply to biological processes and structures and to living things generally.

BIOS Acronym for *b*asic *i*nput *o*utput *s*ystem, the permanent small program that is built into every personal computer and that carries out the operations concerned with the input and output of data, such as the control of the keyboard, the screen display, the disk drives, and the serial and parallel communication connections. The BIOS is usually held on a ROM chip fitted in the machine so that it is always available and booting is possible. BIOS is usually copied from the ROM to the faster RAM soon after switching on.

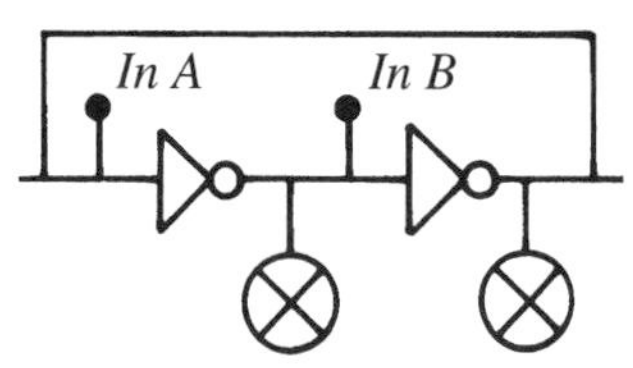

Bistable system

bistable or **latch circuit** An electronic circuit that can have one of two stable states. A bistable usually consists of two linked transistors, one being turned on and the other off, which are capable of being in either of two states. In one state, transistor A is on and B off; in the other, transistor A is off and B on. One state can represent a 1, the other a 0. Bistables can act as volatile memories or buffers and are readily adapted to perform frequency division. If the supply current is turned off, all data held in a collection of bistable memories is lost. Bistables are often called "flip-flops." *See also* astable circuit.

bit Each digit of a number (0 or 1) in binary code is called a bit (from binary digit).

bitmapping An arrangement by which an image for display on a computer screen can be stored. The image is represented in the store by a succession of small-dot picture elements (pixels) arranged in rows and columns. Each pixel is stored as one or more bits of data – a single bit for the presence or absence of light in a simple monochrome system, or several, if shades of gray or colors are required. The number of dots per row or column determines the detail (resolution) of the image, and this is usually expressed in terms of the number of rows and columns. Text can be represented as a bitmapped image, and this differs fundamentally from text represented as ASCII or other alphanumeric code. Bitmapped text, obtained by using a scanner, can be converted to code by means of optical character recognition (OCR) software.

bitter pattern The microscopic appearance of the magnetic domains in a ferromagnetic material revealed by spreading on its surface a colloidal suspension of very fine iron particles.

black body A theoretical body that absorbs all the radiation, of whatever kind, that falls on it. It is also a perfect emitter of radiation. The nearest practical approach to a black body is a hollow vessel with a roughened and blackened inside and a small hole in the surface through which radiation enters and leaves. The energy radiated depends only on the temperature of the cavity.

black body radiation The electromagnetic radiation emitted by a black body. This extends over the whole range of wavelengths and has a characteristic energy distribution with a maximum at a certain wavelength. The peak of this energy varies with the temperature. The higher the temperature, the further the peak moves toward the shorter wavelengths; the lower the temperature, the further it moves toward the long wavelengths.

black box A useful scientific concept or technique in which the units of any system are considered simply as boxes that effect particular functions, or have particular properties, without taking any account of how these functions or properties are achieved.

black hole (in space) An enormous force field emitting no radiation because of its extremely strong gravitational field, which absorbs all matter coming near it.

blast furnace A large tower (approximately 100 feet [30 m] high and 20 feet [6 m] wide) used to extract iron from its ores. Iron ore, coke, and limestone are added from the top and preheated air is blown in through tubes (tuyères) at the base. This causes the coke to burn and leads to several chemical processes, resulting in the reduction of the ore, which settles as a liquid at the base of the tower. A molten slag of calcium silicate floats on this and is removed separately.

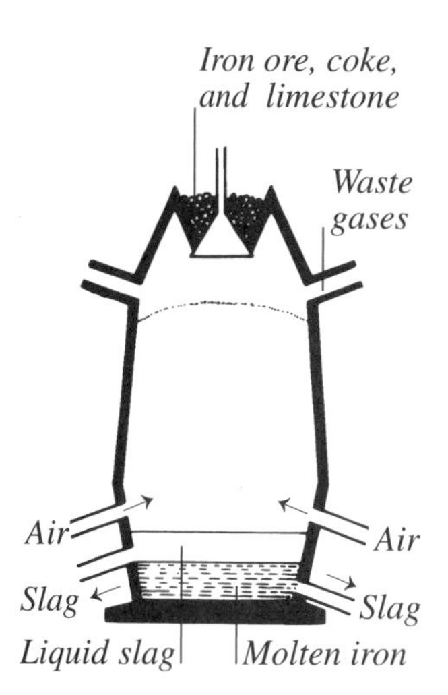

Blast furnace

blind spot The part of the eye where the optic nerve passes through the retina. There are no nerve endings, so it is insensitive to light at this point.

boiler A vessel in which water is heated, producing steam that can drive an engine or turbine, or produce heat.

boiling point The temperature at which a liquid changes to the gas state. It depends on the atmospheric pressure. The boiling point is the temperature, at any particular atmospheric pressure, at which the saturated vapor pressure of the liquid is equal to the atmospheric pressure. This equality allows bubbles to form in the liquid and vapor to escape from it, without a change of temperature, until all the liquid has evaporated. Boiling points of different liquids are given at the standard atmospheric pressure of 760 mm of mercury.

bolometer An electrical instrument used in the measurement of heat radiation. It uses the way in which the resistance of platinum strips changes with temperature.

Boltzmann constant The ratio of the universal gas constant (R) (the constant that appears in the universal gas equation $pV = nRT$) to the Avogadro constant (N_A) (the number of atoms or molecules in one mole of substance). The Boltzmann constant is the gas constant for a single molecule. Its value is 1.380×10^{-23} JK^{-1}.

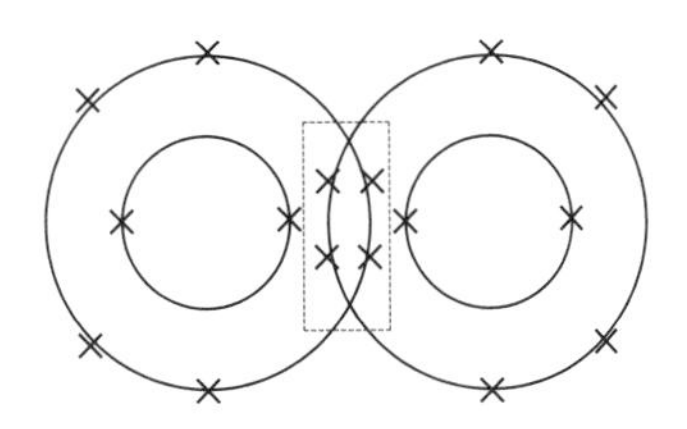

Bond – a double coordinate bond

bond The chemical connection between atoms within a molecule. Bonds are forces and are caused by electrons. Covalent bonds form when two electrons are shared between two atoms (usually between two nonmetallic atoms), one contributed by each atom. Covalent double bonds form when four electrons are shared between the two atoms. Covalent triple bonds form when six electrons are shared between the two atoms. Coordinate bonds are a type of covalent bond that form when one of the atoms supplies both electrons. Ionic bonds (electrovalent or polar bonds) form when atoms form ions and electrons are transferred from one atom to another. The ions are held together by electrostatic attraction. Metallic bonds are chemical bonds between atoms within metals.

Boolean algebra A way of organizing logical operations (AND, OR, NOT) into a system of symbolic logic. Boolean algebra, named for the Irish mathematician George Boole, is the basis of the electronic logic gates without which digital computers would not be possible.

booting The process of bringing a personal computer to the state in which it can accept and run applications programs. A computer cannot function without an operating system program and requires a short program to tell it how to find and load this program into RAM. Such a program is usually held in a nonvolatile read-only memory (ROM) and runs automatically as soon as the machine is switched on. Bringing the machine to this operating state is called booting, or *boot-strapping*. The term derives from the tales of the fantastic adventures of the fictional Baron Munchausen, who, finding himself sinking in a swamp, lifts himself out by his bootstraps.

boson A group of elementary particles (e.g., a photon) that carry force between particles. *Compare* fermions.

boundary layer The very thin layer of fluid molecules surrounding a body immersed in a liquid. These molecules are linked to the molecules of the body and do not move when it moves, relative to the fluid.

Boyle's law The volume of a given mass of gas varies inversely with its pressure at constant temperature. One of the three ideal gas laws.

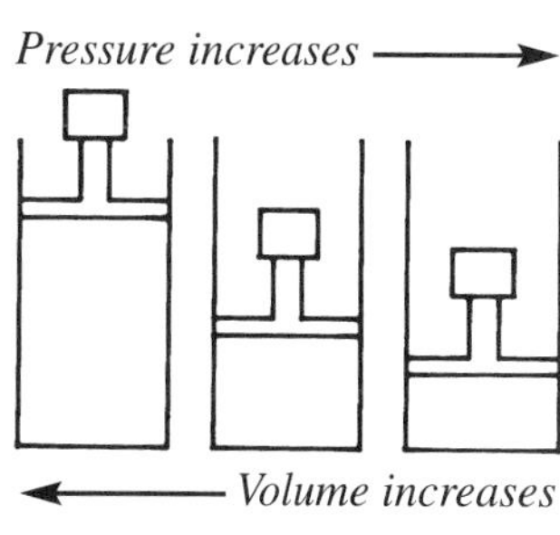

Boyle's law

Brackett series A series of lines in the far infrared of the hydrogen emission spectrum.

branched chain A line of carbon atoms having side groups attached to the chain.

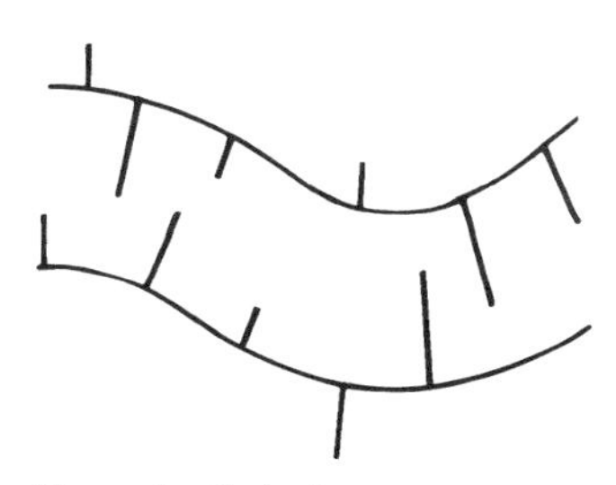
Branched chains

Brewster's law The Brewster angle is the angle of incidence of light where the reflected and refracted beams are at right angles. When the light is incident at this angle, all the reflected light is polarized.

bridge rectifier A component of electrical circuits consisting of diodes connected in a bridge pattern. A different pair of diodes conducts for each direction of AC. It carries out one of the stages of conversion from AC to DC.

brown dwarf A star with a core temperature of less than 107K. It emits mostly infrared radiation.

Brownian motion The constant, fine, random movement observed under the microscope in tiny solid particles immersed in liquid. This is due to the random bombardment of the particles by the moving molecules of the liquid. Brownian movement also occurs in smoke particles suspended in still air. In 1905, Albert Einstein produced an equation that described how unbalanced forces caused by molecular motion would produce the effect. Brownian movement was first observed by the Scottish botanist Robert Brown while examining pollen grains.

bubble chamber A method of demonstrating the movement of subatomic particles of ionizing radiation. The chamber contains liquid hydrogen kept at a very low temperature, just above its boiling point but under high pressure, so as to prevent vaporization. Immediately before the radiation is to be observed, the pressure in the chamber is suddenly reduced so that the liquid boils. The moving particles then leave a wake of tiny bubbles that can easily be photographed. The device was invented in 1952 by American physicist Donald Glaser.

bug In computing, a bug is an error accidentally introduced into a program so that it either fails to implement the intention of the programmer or fails to operate at all. Debugging often takes longer than writing the program, but software (debuggers) exists to facilitate the process of detecting and eliminating bugs. The so-called Millennium bug was not a bug but the result of a failure of many programmers to anticipate the effects on the operation of their software of their practice of using only two figures to represent a year that can only be fully specified with four figures.

bulk modulus If a solid body is compressed by an external force and returns to its original volume when the force is removed, the constant of elasticity is called the bulk modulus. *See* elasticity.

Bunsen A burner used in the laboratory. It burns a variable mixture of gas and air, the proportions of which can be changed by manipulating the air hole on the side of the burner.

Bunsen burner

burette A long, graduated glass tube with a tap at the lower end. It is used to accurately measure out a volume of liquid.

bus A system of electrical connections or wires that carry related data or instructions between the various parts of a computer. Ideally, the number of lines in a bus is equal to the number of bits in the word size the machine is capable of managing. Thus a 16-bit bus would most efficiently have 16 lines. Computers now commonly use 32- or even 64-bit words, and as multichannel buses are expensive and difficult to miniaturize, various expedients such as multiplexing are often adopted. Also known as a highway.

byte A collection of binary digits, now universally numbering eight on modern systems and usually coding for a number or a character. Each of the 255 characters in the extended ASCII code can be represented by a single byte. A kilobyte is not 1,000 bytes, but 1,024, and a megabyte is not 1,000,000 bytes, but 1,048,576. These are the decimal representations of the binary numbers nearest to the nominal number. A gigabyte is something more than a billion bytes.

C A widely used high-level and compact programing language developed in the mid-1970s and popular with programmers, especially for use in personal computers. C can be used for a variety of purposes, and, because it is close to machine language, produces highly efficient code. A development of C, C++, provides programmers with the ability to define the type of a data structure and also the kind of operations or functions that can be applied to the data structure. The data structure is thus regarded as an object that includes both data and functions. This is a highly modular and efficient method of working, known as object-oriented programming (OOP).

cache memory A high-speed computer memory, common in high-performance systems and independent of both the main random access memory (RAM) and the hard disk but sometimes a part of the RAM set aside. Cache memory may be situated between the hard disk memory and the central processing unit. Because access to it is much faster than access to the hard disk, it can, if used to hold material to which reference is repeatedly made, greatly speed up the operation of the machine.

CAD Acronym for *c*omputer-*a*ided *d*esign. CAD programs allow designers to implement designs in graphic form without having to make actual physical models. These designs can be viewed from various directions, zoomed in and out, dimensionally altered, rendered, colored, modified in many ways, and printed out in full color. When one design value is changed, the computer will automatically change other dependent values. Many design decisions can thus be made very much more cheaply and quickly than by earlier methods.

CAD/CAM Acronym for *c*omputer-*a*ided *d*esign/*c*omputer-*a*ided *m*anufacture, a system in which the software, after assisting with the design, can be used to control the manufacturing processes.

calorimeter A vessel used in heat measurements. It is made of a good conductor, so that it will be at the same temperature as its contents, and its surface is polished to minimize heat loss by radiation. To minimize heat loss by conduction and convection, it is placed within a larger lagged container.

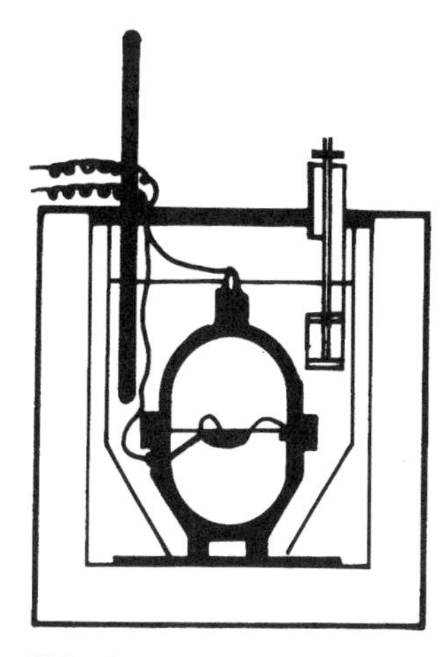

Calorimeter

candela The SI unit of luminous intensity.

capacitance A measure of the ability of an object to store electrical charge as its potential rises. The capacitance of a conductor is a ratio of its charge to its potential.

capacitor A device, consisting of a pair of metal plates separated by an insulator (dielectric), that can store an electric charge. A wide range of capacitors, with capacitance ranging from picofarads to farads, is used in electronics and electrical engineering.

capillary A very fine hairlike tube. A capillary tube has a narrow bore and thick walls, like a thermometer tube.

carbon chain *See* chain.

carbon cycle The processes by which carbon is circulated around the Earth. Carbon dioxide is released into the atmosphere by the decay of animal and vegetable remains, the breathing of animals, and the burning of fossil fuels. Carbon dioxide is used by green plants in photosynthesis (it also dissolves in water), from which it is released by decay.

carbon dating (radiocarbon dating) The way in which the age of previously living animal or vegetable life can be determined. Carbon is present in the atmosphere and in all living tissue in a mixture (the proportions of which are constant while the tissue is living) of isotopes, one of which, carbon-14, is radioactive with a half-life of 5,730 years. When the tissue (animal or vegetable) dies, the proportion of carbon-14 decreases as it undergoes radioactive decay. The age of a sample of dead material can thus be measured by measuring the radioactivity of the sample.

Carnot cycle A closed cycle describing pressure and volume changes in a system that operates between different temperatures. If these changes are described for an ideal heat engine, where heat is converted to work, the maximum efficiency depends solely on the operating temperatures between which it works. The engine absorbs heat at the higher temperature and rejects waste heat at the lower temperature.

carrier wave The electromagnetic wave of regular frequency, emitted from a transmitter, on which a second wave is superimposed by altering either its magnitude (amplitude) or the number of its cycles per second (its frequency). This superimposition on the plain wave is called modulation. All radio communication is mediated by carrier waves, which may be amplitude modulated (AM), frequency modulated (FM), or pulse code modulated (PCM). The latter two are much less susceptible to interference than the former. FM is preferred for public broadcasting, PCM for data transfer. *See also* frequency modulation, amplitude modulation.

Cartesian coordinates Any system of locating a point on a plane by specifying the distance along two axes (usually a horizontal X axis and a vertical Y axis), or a point in space by specifying the distance along three axes (X,Y, and Z). The axes intersect at the origin. Below the origin the values on the Y axis are negative, and to the left of the origin those on the X axis are negative. The system is named after its originator, the French philosopher and mathematician René Descartes.

catalyst A substance that increases the rate of a chemical reaction. It takes part in the reaction but remains chemically unchanged by it. Enzymes are the organic catalysts present in animals and plants.

cathode (1) A negative electrode from which electrons (negatively charged particles) emerge. A cation is an ion that is attracted to the cathode and is thus positively charged. An anion is negatively charged.
(2) The electrode carrying the negative charge in a solution undergoing electrolysis.

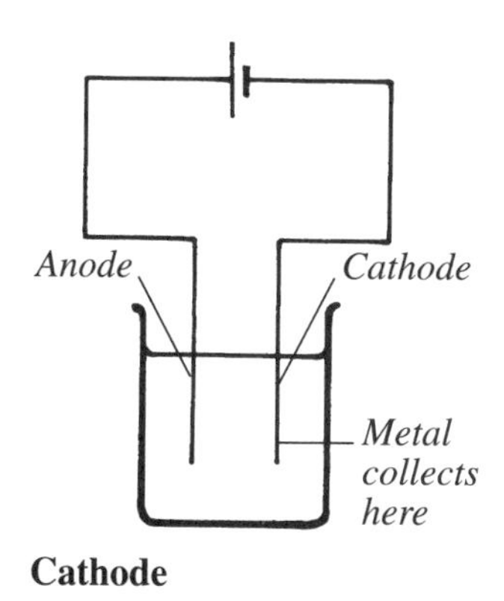

Cathode

cathode-ray oscilloscope (CRO) An instrument used to display electrical signals on a phosphorescent screen. It is particularly useful in the display of radio and audio signals that have rapidly changing voltages.

cathode rays The stream of electrons emitted from the cathode in a vacuum tube.

cathode-ray tube A widely used electronic display device consisting of an evacuated glass tube with an expanded and flattened end, the screen, coated on the inside with phosphors that emit visible light when struck by a beam of electrons. The electron beam is produced by a heated filament at the narrow end of the tube, and the negatively charged electrons are accelerated toward the screen by a strong positive voltage applied to it. The electron beam, being charged, can easily be focused and deflected by an electric or a magnetic field. Two electronic oscillators, the time bases, provide fluctuating voltages that cause the electron beam to scan rapidly horizontally and more slowly vertically, thus covering the whole screen in a fraction of a second. While so doing, the beam is repeatedly interrupted so that no light is produced. The image on the screen, whether textual or graphic, is built up by the uninterrupted beam. Cathode-ray tubes are used in television, personal computers, oscilloscopes, and radar displays, as well as various communication systems. They are gradually being displaced by more compact and flat display systems, such as liquid-crystal screens.

cation An ion having positive charge, which is attracted by the negatively charged electrode, the cathode, during electrolysis.

caustic curve A curve or surface that is caused when parallel light rays strike a concave mirror. The rays are reflected in such a way that they intersect to form a curve. Such a curve is seen on the surface of a liquid in a cup.

CD-ROM A standard compact disk used as a large-capacity data store for textual or graphic information. CD-ROMs are identical to music compact disks in which the information is stored in digital form as microscopic pits cut in a reflecting surface. A CD-ROM can store about 600 megabytes of information – about 100 million words of text – and this can be substantially increased by various compression techniques. It thus represents an efficient and compact form of information storage that is being widely exploited as a substitute for the printed word. CD-ROM drives are standard peripherals on personal computers and are addressed in exactly the same way as a standard hard drive. With suitable search and retrieval software, information access is much quicker than from printed books.

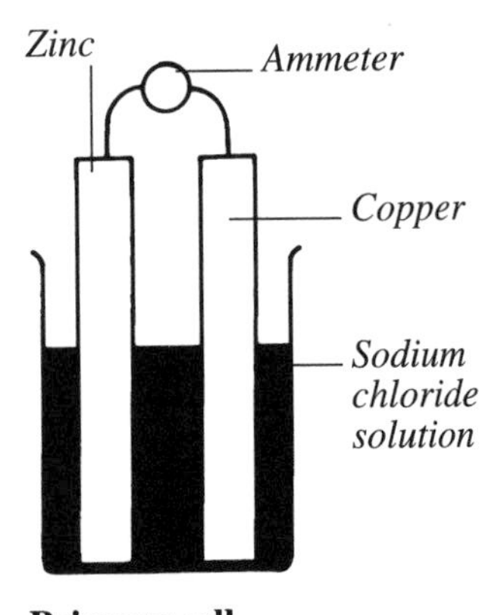

Primary cell

cell A vessel, used either to produce electricity or to perform electrolysis, containing an electrolyte in which two electrodes are dipped. There are three main types of cell:
(1) primary cells, which produce electricity by chemical action (usually irreversible);
(2) secondary cells, which can be "charged" by passing electricity through in a direction opposite to the discharge (this reverses the chemical action, which produces electricity);
(3) electrolytic cells in which electrolysis takes place.

Celsius A scale of temperature that has 100 divisions between the lower fixed point (the melting point of pure ice) and the upper fixed point (the boiling point of pure water). The degree is the same size as the kelvin in the absolute temperature scale (*see* absolute zero). This scale is identical to the centigrade scale, but the name was changed in 1948 to avoid confusion with the grade (one-hundredth part of a right angle) and to honor the Swedish astronomer Anders Celsius, who invented a similar, but inverted, scale in 1742 and whose surname conveniently begins with a C.

center of mass The point at which the whole of the mass of a body can be considered to be concentrated and to act for gravitational or other purposes. The center of mass is the same as the center of gravity.

centrifugal force An imaginary entity widely, but wrongly, believed to exist. It is deemed to act outwards as a balance for the centripetal force – the force acting inwards that causes a body to move in a circular path. It is not centrifugal force that causes a stone swung on a string to move off at a tangent when the string breaks; it is the sudden loss of the centripetal force that allows the stone to continue to conform to Newton's first law of motion and to proceed in a straight line.

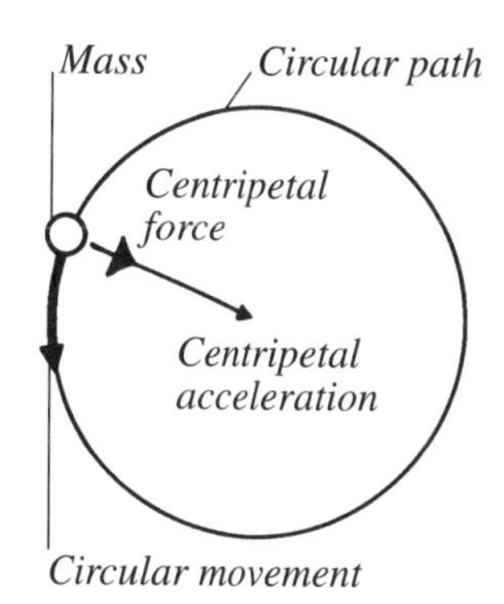

Centripetal force

centripetal force The force acting radially inwards on a body that is rotating in a circle, constraining it to move in a curved path.

Cerenkov radiation The electromagnetic radiation effect, analogous to the sonic boom in acoustics, in which a bluish light is emitted when a beam of high-energy particles passes though a transparent medium at a speed greater than the speed of light in that medium. Light travels more slowly through transparent media than in a vacuum.

CERN Abbreviation for Conseil Européen pour la Recherche Nucléaire, the European organization for atomic research, based in Geneva. This is one of the major high-energy physics research centers in the world.

c.g.s. units A system of units based on the centimeter, the gram, and the second, which never proved comfortably consistent with the equivalent units for heat and electricity and which, for many scientific purposes, has been superseded by the Système International (SI) system of units.

chain A line of carbon atoms in a molecule. These chains can be thousands of atoms long in polymers. Molecules can consist of a straight or branched chain.

chain dial balance A balance in which the use of small weights is replaced by a chain attached to one end of the beam and at the other to a vertical scale.

chain length A measure of the number of atoms linked to form a chain.

chain reaction A reaction in which one event leads to a second, and so on. It is often used to describe a nuclear reaction in which energy is released constantly because neutrons emitted by the fission of an atomic nucleus proceed to cause further fissions, which in turn emit more neutrons.

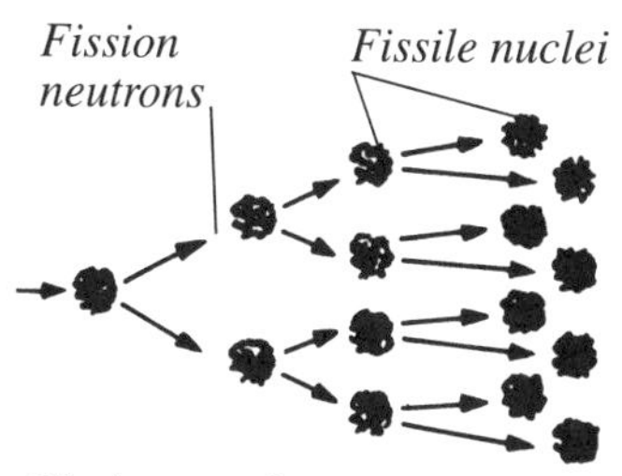

Chain reaction

chamber process One of the processes used for sulfuric acid production. Sulfur dioxide, oxygen, and nitrogen dioxide react within a large,

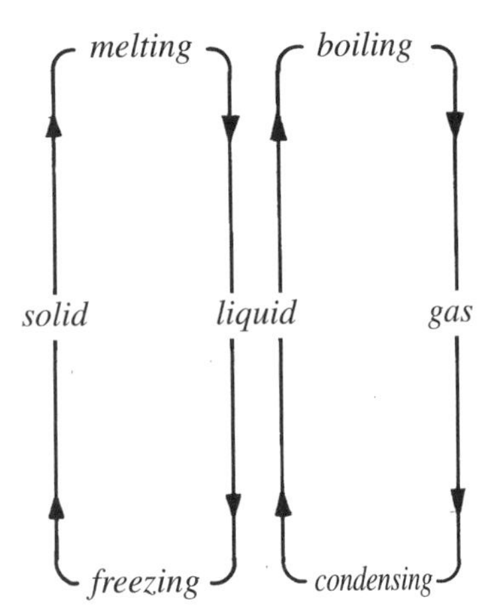

Change of state

lead-sheathed brick tower. Sulfuric acid forms as fine droplets that fall to the base of the tower.

change of state The physical process where matter moves from one state to another. Examples of such changes are melting, evaporation, boiling, condensation, freezing, crystalization, and sublimation. A change of state is associated with energy changes.

charge (electric charge) This is the property of certain atomic and subatomic particles that causes forces of attraction and repulsion between each other. The electron has the smallest unit of negative charge, the proton an equal amount of positive charge. Like charges repel; opposite charges attract.

charge density The distribution of charge over the surface of an object. It is spread evenly over the surface of a curved object. On a pointed object, the charge density will be greatest at the tip.

charge-couple device (CCD) (1) A sensor capturing images in astronomical telescopes and camcorders. They consist of a very large number of very small picture element detectors.
(2) A device for charge storage memory.

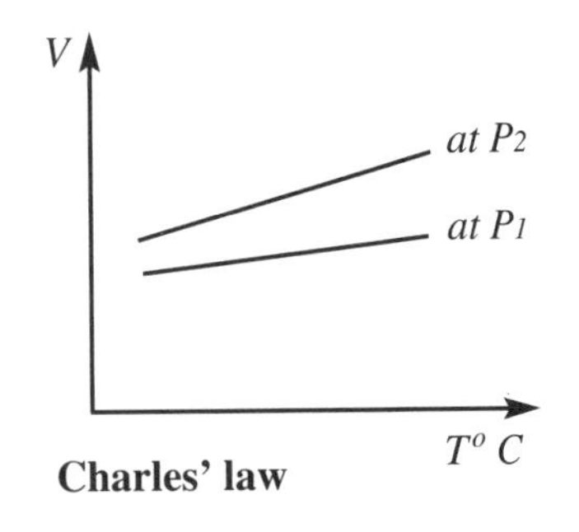

Charles' law

Charles' law The volume of a fixed mass of gas at constant pressure is directly proportional to its temperature.

chemical energy The energy stored in the bonds between atoms and molecules that is released during a chemical reaction.

chlor-alkali cell Hydrogen, sodium hydroxide, and chlorine are produced in a chlor-alkali cell by the electrolysis of brine using a flowing mercury cathode (the Castner-Kellner cell).

choke An electronic circuit element, consisting of an inductor with a high inductance but a low electrical resistance, that offers a high impedance to alternating or fluctuating current but a low impedance to direct current. Chokes are useful in smoothing power supplies after rectification of alternating current, but the development of cheap, very high-capacity electrolytic capacitors has made the more expensive chokes less necessary.

Chemical energy

chromatic aberration The colored rainbowlike bands that occur around an image formed by a lens because light of different wavelengths is bent to different degrees as it passes through it. Chromatic aberration can be reduced by designing a lens using two glasses of different refractive index cemented together.

chromatography A way of separating and identifying mixtures of solutes in a solution. The method depends on the affinity of the different solutes in the mixture for the medium through which the solution moves.

chromosphere A very thin region of low density in the lower atmosphere of the Sun, seen as a rose-colored ring during a total solar eclipse. This area contains some relatively cool gases that absorb wavelengths of light, causing dark absorption lines in the solar spectrum called Fraunhofer lines.

chronometer A very accurate clock, regulated by a balance wheel and hair, used in navigation to determine longitude.

circuit board A thin, rigid board of insulating material on which a very thin layer of copper is deposited, usually on both sides, and chemically cut into thin interconnecting strips by which microchips, resistors, capacitors, quartz crystals, and other electronic components are interconnected. The circuit board nearly always has a multi-way edge connector or connectors by which it can be plugged into other circuits or motherboards, or to which power-supply connectors can be attached. System failure can thus rapidly be remedied by removing and replacing the faulty board with a new component.

circular measure A way of measuring an angle by considering it as the angle formed at the center of a circle by the lines running to the center from the ends of an arc of the circumference. The unit of circular measure is the radian, the angle subtended at the center of a circle by an arc equal to the radius of the circle. The circumference of a circle is 2π times the radius, so the whole circumference will subtend an angle of $2\pi r$ divided by r radians, which is 2π radians. Thus 360° is equal to 2π radians, and 1 radian is equal to 57.296°.

Equal to radius

1 rad

Circular measure

circumpolar orbit A circular orbit centered on the Earth's center, passing over both North and South Poles.

classical physics Theoretical physics before the advent of the quantum theory and relativity in 1900 and 1905, respectively. Classical physics was based primarily on Newtonian ideas and on Clerk Maxwell's equations for electromagnetism. It is still entirely valid for most observable phenomena involving bodies containing large numbers of atoms and speeds well below that of light.

clipboard A small segment of memory or a file in which data copied or cut from one place are stored temporarily until they can be pasted into another place. The Windows 95 and 98 operating systems incorporate clipboards that can be used in most applications programs. These allow word processors, for instance, to cut a block of text from one document on the screen, store it in the clipboard, and then paste it into another document, or another part of the same document.

clocked logic Logic gates used in a circuit that generate pulses at precisely controlled intervals. All personal computers use clocked logic based on a crystal-controlled pulse generator circuit running at fixed speeds that may be as high as 1 billion pulses per second (1,000 megahertz).

close-packing structures Atoms packed together in a structure so that the closest separation of centers of atoms equals the diameter of the atom.

cloud Condensed water vapor floating in a mass in the air. Clouds form when air cools and the water vapor contained in the air condenses into tiny drops of water or ice. The appearance of clouds and their names depend on their altitude and shape. High-level (40,000 feet [10,000 m] altitude) cirrus clouds appear wispy and consist of minute specks of ice. Medium-level (9,000–23,000 feet [3,000–7,500 m]) cloud names have the prefix "alto." Low-level (below 6,500 feet [1800 m]) cloud names have the prefix "stratus" and have a layered structure. Nimbus clouds are shapeless, gray rainclouds. Cumulus clouds are dome-shaped with a flat base at a low level (4,000 feet [1,400 m]) extending to about 6,500 feet (1,800 m).

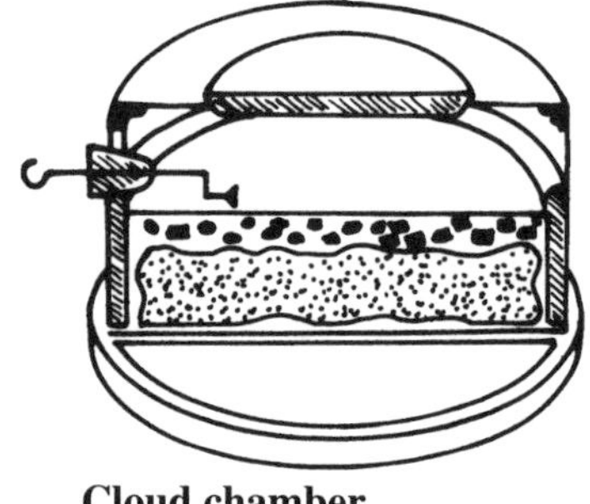

Cloud chamber

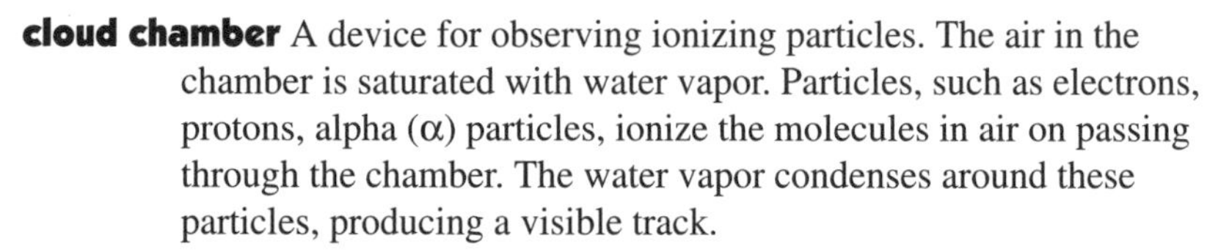

cloud chamber A device for observing ionizing particles. The air in the chamber is saturated with water vapor. Particles, such as electrons, protons, alpha (α) particles, ionize the molecules in air on passing through the chamber. The water vapor condenses around these particles, producing a visible track.

CMOS chip Abbreviation for complementary metal oxide semiconductor chip. These have very low-power consumption and can operate on low-supply voltages.

coaxial cable An electric cable consisting of a central conductor separated by insulation from a second, cylindrical conductor that surrounds it on the same axis. Coaxial cable is used to carry high-frequency electrical signals, as it is relatively immune to external interference and carries signals with minimal loss. Televison aerials (antennas) are usually connected to their receivers by coaxial cable.

coefficient of friction A measure of the ease with which two surfaces move relative to each other. $P = \mu R$, where P is the force with which the object is pulled, μ is the coefficient of friction, and R is the force at right angles between the surfaces (the normal reaction). Its value is less for moving surfaces than for static ones.

coefficient of thermal conductivity A measure of a material's capacity to conduct heat energy.

coefficient of viscosity A measure of the viscosity (η) of a fluid.

collision theory Particles of a reactant must collide if they are to bond. There must be sufficient energy in the collision to cause the reorganization of bonds needed in a reaction. The particles must be aligned correctly when they collide, or new bonds may not form.

colloid A substance made of very small particles, whose size (1–100 nm) is between those in a suspension and those in a solution.

color The visible part of the spectrum of electromagnetic radiation, although seen as white light, consists of different wavelengths, each having a different color. The eye senses different colors when particular wavelengths of light strike the retina. Primary colors in photographic reproduction are red, green, and blue. When they are mixed, they form white light. If light of any two primary colors is mixed, a secondary color is formed. Red and blue light form magenta light. Blue and green light form cyan light. Red and green light form yellow light.

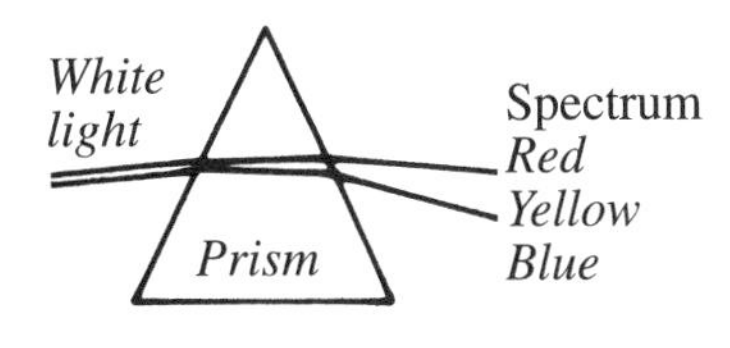

Color

combinational logic design A number of logic gates connected together.

combustion The chemical term for burning, usually in oxygen.

combustion tube A closed tube in which combustion takes place.

command.com The file that contains all the internal commands for the standard Microsoft disk operating system (MS-DOS) on which the majority of personal computers, including those running Windows 95 and Windows 98, operate.

commutation A term used in Boolean algebra when the same result is obtained, irrespective of the order of the arguments.

commutator A device that is used both to change the direction of current flowing in the armature coil of a DC motor and in a generator to convert AC/DC. A simple commutator is made of a split copper ring, the two halves being insulated from each other. Each half is connected to the armature coil and rotates with the coil. Carbon brushes make connections to external circuits.

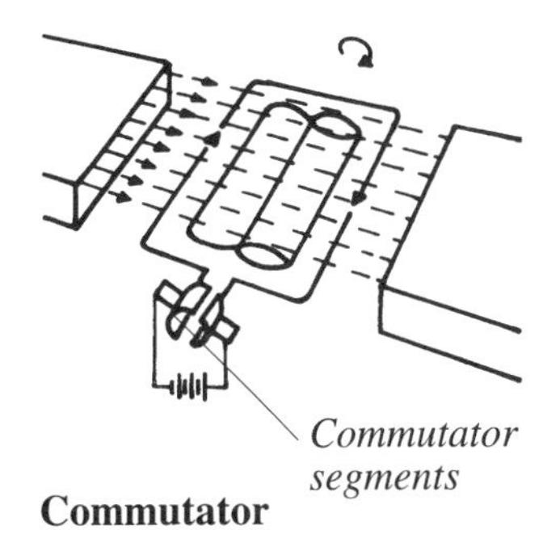

Commutator

compound Two or more elements chemically joined together.

compression The result of applying force or pressure to a body so that it becomes smaller or shorter.

concave Curving inwards, like the interior of a sphere. A concave lens is thinner at its center than at its periphery and causes incoming light rays to diverge. A concave mirror is a converging mirror.

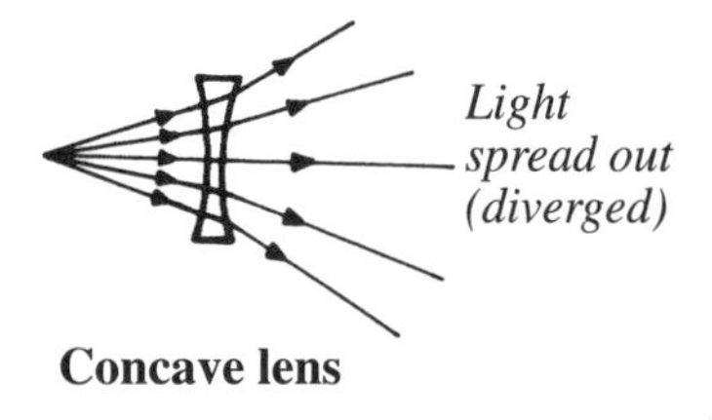

Concave lens

concentration A measure of the quantity of solute contained in a solution at a given temperature.

condensation The process by which a liquid forms from its vapor.

condensation reaction The joining together of two or more molecules with the elimination of a small molecule (usually water).

condenser (1) An apparatus in which a vapor is converted to a liquid. In a condenser (Liebig condenser), the tube through which the vapor flows is surrounded by a jacket in which water flows.
(2) A combination of optical instruments (mirror, lens) that concentrates light.

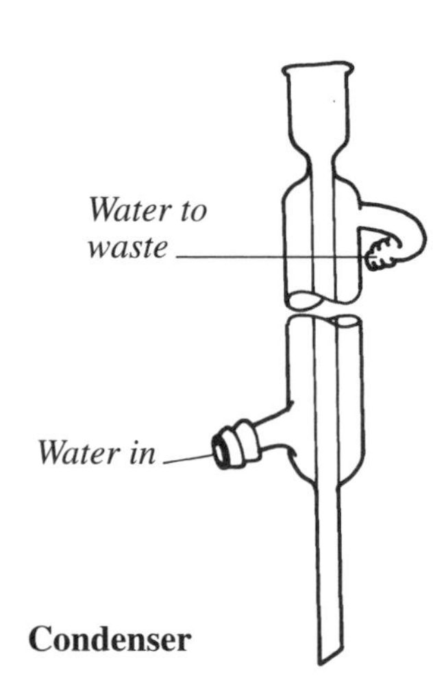

Condenser

conductance The ability of a conductor to conduct an electrical current.

conduction (1) (*electrical*) The movement of free electrons from atom to atom in a metallic conductor that transfers electrical energy. The current (flow of charge per second) depends on the circuit's resistance (Ohm's law).
(2) (*thermal*) *See* thermal conduction.

conductivity (1) (*electrical*) A measure of the ease with which a substance conducts electricity.
(2) (*thermal*) *See* thermal conductivity.

conductor A material that is able to conduct heat and electricity.

config.sys The file that organizes (configures) the way a personal computer using the MS-DOS or other similar operating systems is set up so that it can proceed to run various applications programs.

conservation of energy, law of or **first law of thermodynamics** In a closed system, energy remains constant. Energy can be neither created nor destroyed.

conservation of momentum The total momentum of colliding bodies prior to impact equals their total momentum after impact has taken place, in the absence of external forces.

contact process The industrial process used to manufacture sulfuric acid. It uses iron pyrites.

control rod Rod used in a nuclear reactor to maintain the reaction at the correct level. The rods are made of boron, which absorbs the neutrons that would otherwise cause further fissions. They are raised or lowered into the reactor core depending on the level of reaction required.

convection The transfer of heat energy in fluids by motion of the fluid. Molecules in contact with a heat source gain energy, move faster, move apart, and rise; colder, denser molecules take their place, causing convection currents.

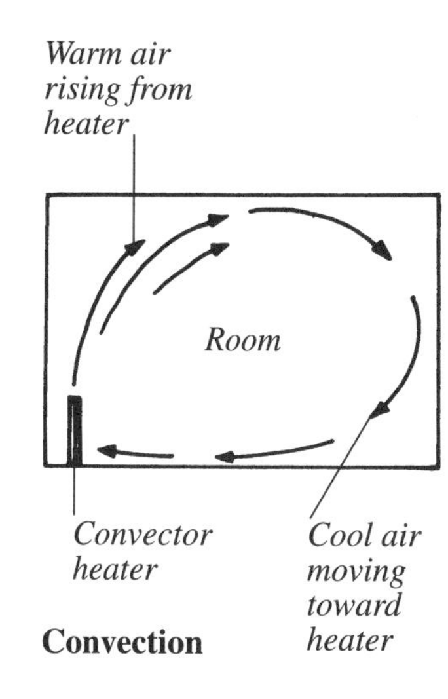

Convection

converging rays Narrow beams of radiation that move toward the same point.

convex Curving outward, like the outside of a sphere. A convex lens is thicker at its center than its periphery. It causes rays to converge. A convex mirror causes rays to diverge, magnifying the image.

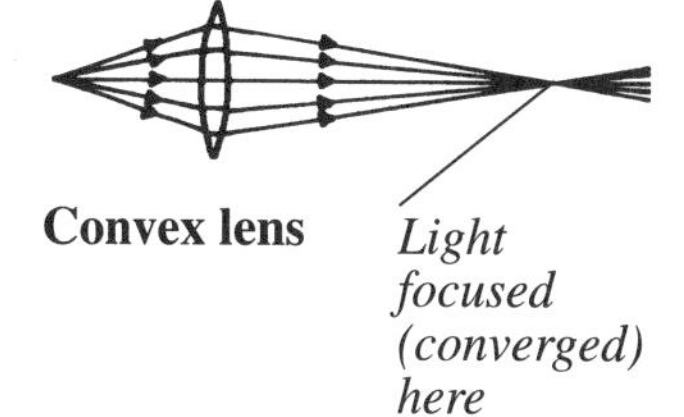

Convex lens

coordinate The set of numbers used to locate a point with reference to a system of axes.

coordinate bond *See* bond.

core charge In a molecule having covalent bonds, such as water, where the oxygen nucleus is more massive than the hydrogen nucleus, electrons in the shared pairs are closer to the oxygen nucleus because of its larger attracting charge than the electrons in the lone pairs.

cornea The transparent convex membrane covering the front of the eye.

corrosometer An application of the Wheatstone bridge, used to measure a material's susceptibility to corrosion.

coulomb The charge carried by an electric current of one ampere in one second. The SI unit of electric charge.

Coulomb's law The forces of attraction or repulsion between two small charged bodies is directly proportional to the product of the charges and the distance between them.

couple Two equal and parallel forces that act together in opposite directions, tending to turn an object.

covalency The number of covalent bonds an atom is able to make when forming a molecule.

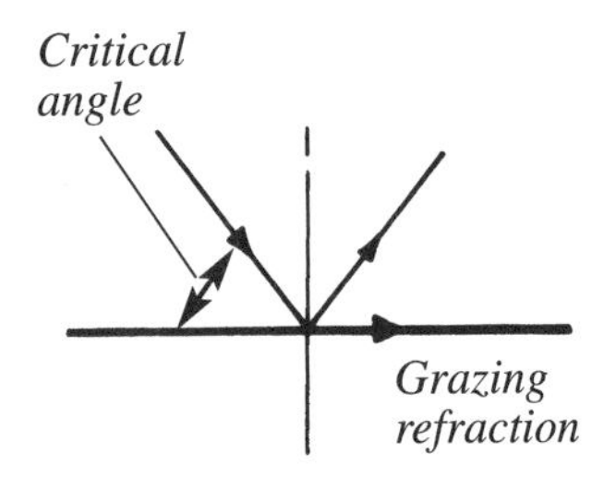

Critical angle

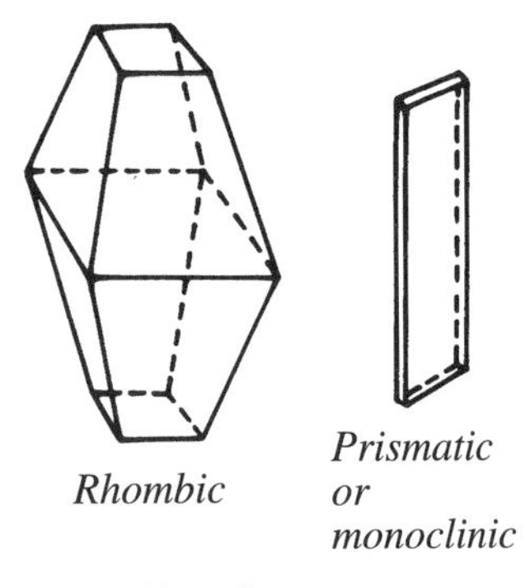

Crystaline shapes

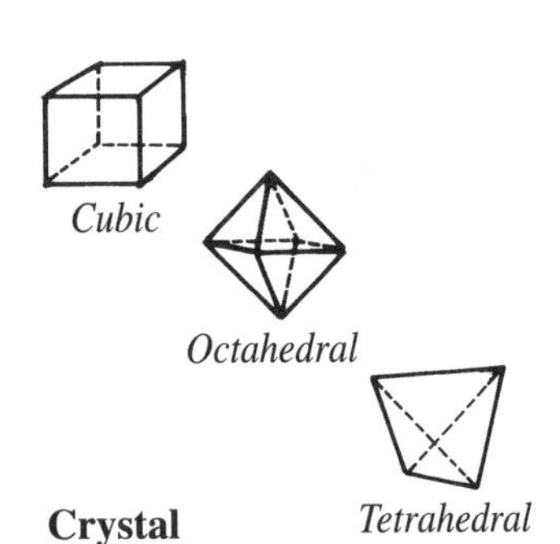

Crystal

covalent bond *See* bond.

creep The retention of a change in shape in a material after the stress causing it has been removed.

critical angle The smallest angle of incidence of a light ray passing from a dense to a less dense material, at which light only just reappears (the angle of refraction is 90°). This is called grazing refraction. At a larger angle of incidence than the critical angle, total internal reflection occurs.

critical temperature The temperature above which a gas cannot be compressed into a liquid.

cross-linking Chemical bonds between adjacent polymer molecules.

crucible A vessel constructed from heat-resistant material (porcelain) that is used to heat substances to high temperatures.

crystalline Composed of crystals.

crystal structure The orderly geometric arrangement, or lattice, of atoms, molecules, or ions in a structure that has a particular regular three-dimensional structure. There are several basic shapes taken by a crystal lattice, depending on the component particles. Shapes can be cubic, tetragonal, rhombic, hexagonal, trigonal, monoclinic, or triclinic. In addition, they can have close-packed structures, in which the lattice is said to be face-centered, or more loosely packed, in which case the lattice is body-centered.

cubic expansivity The cubic expansivity of a substance is approximately three times its linear expansivity.

cubic relationship The volume of a cube that is proportional to the cube of its length.

current or **electric current** A flow of electric charges moving through a conductor.

cyclotron An electromagnetic accelerator that produces an intense beam of charged high-energy particles, such as protons and electrons.

daisychaining A method of connecting a series of devices to a single controlling circuit using a single cable interrupted by each device. Daisychaining is one of the most valuable features of the small computer systems interface (SCSI – pronounced “scuzzy”) used to connect external disk drives and other peripheral devices to a personal computer.

damping The reduction in the amplitude of any oscillating or vibrating system as a result of the dissipation of energy through friction or in any other way.

dark matter Matter in the universe whose presence can only be deduced. Calculations of the mass of the universe exceed by several times the total estimated mass of the visible stars. The difference might be accounted for by black holes, nonluminous dwarf stars, or other unidentified matter.

database An electronic filing system, a collection of computer-accessible information on any subject, coded and stored on magnetic, optical, or other media, and so arranged that any item can be quickly retrieved for use. Minimal items of information are called fields; groups of these are called records; and collections of records are called files. A database management system is a collection of programs that allows storage, organization, amendment, and retrieval of database information. Database organization can be simple but rigid (flatfile), hierarchical, or complex but versatile (relational).

data communications The process of transferring information from one point to another. In practice, the phrase tends to be used in a narrower sense and to imply the interchange of digitally coded information, whether alphanumeric or graphic. The world now has an extensive network of data communications and these are becoming ever more complex and sophisticated. An important development in recent years is packet switching.

data compression One of several methods of treating coded data so that, without any loss of information, the material can be stored using a reduced number of binary digits. Textual material contains a great deal of redundancy and can be reduced by a factor of four or more. Other forms of data, such as graphics, cannot be so efficiently compressed. The capacity of data storage systems, such as hard disks in personal computers, can readily be doubled by purely software compression methods.

DC Abbreviation for direct current.

DC motor An electric motor that uses direct electric current.

de Broglie relationship The identity between particles and waves that forms the basis of wave mechanics. This identity was first described by the French physicist Prince (later Duke) Louis de Broglie.

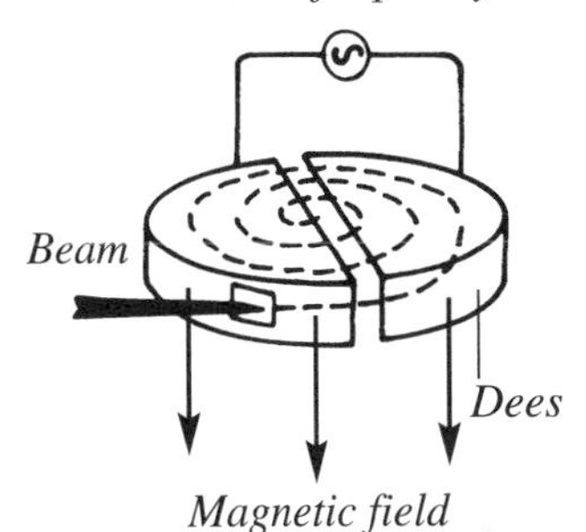

Cyclotron

de Broglie wavelength The length of the wave associated with a moving particle, such as an electron. The wavelength is equal to Planck's constant, divided by the product of the mass of the particle and its velocity (the momentum of the body).

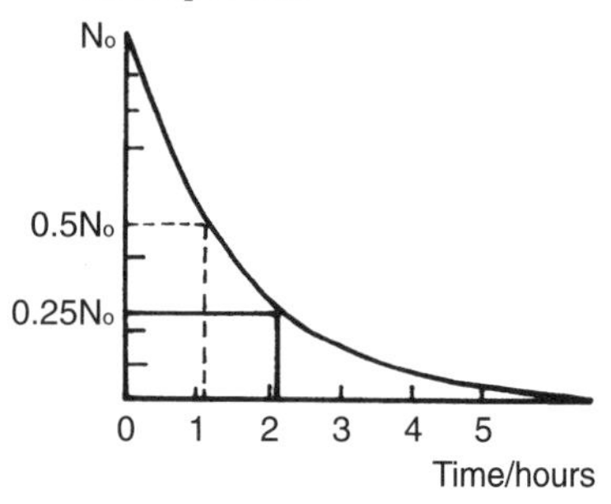

Decay

decay (*radioactive*) The spontaneous breakdown of the nuclei of a radioactive element into another element, with the emission of particles or photons. In the process, the atomic number may rise or fall. The half-life of the element is the time taken for half of the element to decay. If the original amount of the element is x, then the time taken for x to reach x/2 is the same as for x/2 to reach x/4, and so on. *See also* alpha particle, beta particle, gamma radiation.

deceleration If the velocity of a body decreases, its acceleration is negative and is termed *deceleration*.

decibel (dB) A unit of comparison between an arbitrary standard power level and any other level. It is not, as is often thought, a unit of sound intensity. The decibel is commonly used to compare sound intensities or electrical signals. In sound levels, one decibel is about the smallest difference perceptible. It has a logarithm scale, which is convenient because of the very wide range of comparisons commonly made. The scale of human perception of audibility, for instance, ranges from 1 to 1,012. The decibel is equal to one-tenth of a bel. The latter is too large for practical purposes and is seldom, if ever, used.

decision gate An electronic logic circuit that discriminates between true and false logical relationships, giving an appropriate output, such as a 1 for true and a 0 for false. Discrimination might be as to whether it is true or false that one number is equal to another, or greater or less then another.

decoder/demultiplexer An electronic switching circuit that separates and directs to their appropriate outputs the sequential packets of data in a multiplexed signal.

deep inelastic collisions The brief interaction between two colliding atomic nuclei while their surfaces are overlapping. Energy and mass flow from one to the other, and neutrons and protons are exchanged.

degaussing The process of removing the magnetization of an object that has become magnetized. This is commonly done by applying a rapidly diminishing alternating current to a coil so as to produce an alternating and diminishing magnetic field.

degrees of freedom The number of different mutually perpendicular planes in which a system can move, or, more generally, the number of independent parameters required to specify the configuration of a system.

dehydration A chemical reaction to remove a water molecule from a compound. A dehydrating agent is used to remove water from other substances.

delayed neutron emission The spontaneous emission of a neutron from the nucleus of an atom of a radioactive element as a result of energy derived from prior radioactive decay. In a nuclear reactor using uranium-235, 0.7% of emitted neutrons are delayed. This is fortunate, as prompt neutrons take only about one-thousandth of a second to be absorbed, produce fission, and generate further neutron emission. Were if not for the delayed neutrons, there would be very little time to move in the control rods in the event of a threatened runaway chain reaction. The delayed neutrons increase the safe response time from milliseconds to a few seconds.

delay line A component that produces a lag in the time of arrival of a signal in an electronic circuit, compared with the signal in a conductor that bypasses the delay line. Delay lines were used in some early digital computers but are no longer appropriate for this purpose.

deliquescence The way in which a solid substance absorbs water from the atmosphere. The process can continue until the substance passes into solution.

delocalized electron Each atom in a metal has one or more outer electrons that are free to move between atoms. These are delocalized electrons.

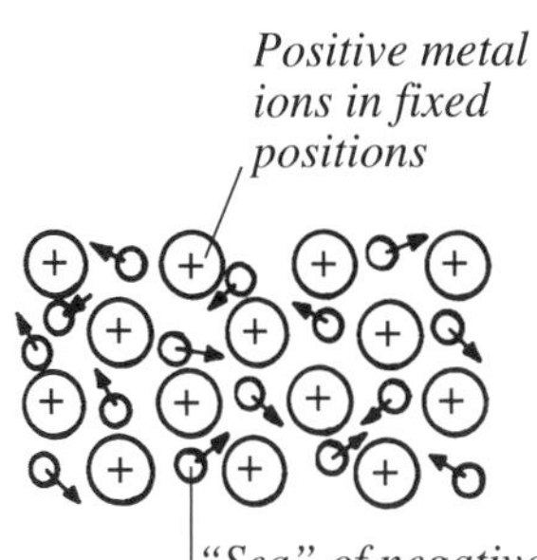

Delocalized electrons

demodulation The retrieval of the information content of a modulated carrier wave. Demodulation occurs in every radio receiver.

densitometry The process of measuring the optical density of a semitransparent medium such as a photographic film. Light is shone through the medium and its intensity is then measured by a photoelectric cell and meter.

density Mass per unit volume of a substance. In SI units, kg/m^{-3}. Density is not the same as specific gravity. This is actually the ratio of the density of a material to that of a standard, such as water, at a standard temperature, which is taken as 1. Because nearly all materials change volume with changes in temperature, density is a function of temperature. In the case of gases, it is also a function of

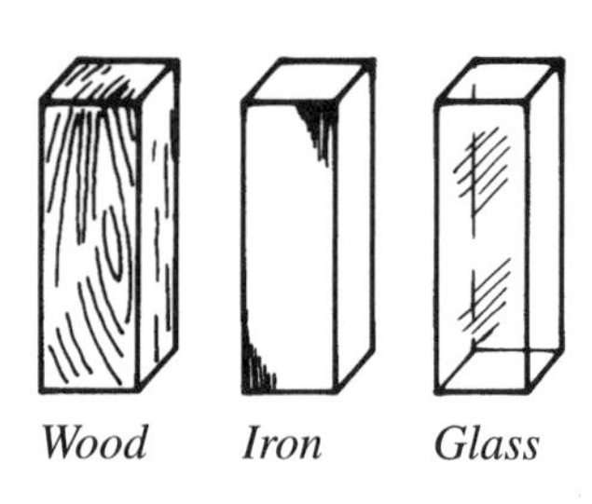

Density *(see page 41)*

pressure. Density of a uniform material is found by determining the volume of a sample and then weighing it. Volume can be determined by immersing the body in a liquid in a graduated container and noting the rise in the level. A better way of determining volume is to weigh the body in air and then to weigh it immersed in a liquid of known density. The volume is equal to the reduction in weight in the liquid, divided by the density of the liquid.

depletion layer A zone in a semiconductor with fewer than the usual number of mobile charge carriers (electrons and holes). Depletion layers form at the interface between p-type and n-type material. Semiconductor function depends on this fact.

depolarization The use of a material to prevent the positive pole of a primary electric cell from becoming polarized by the accumulation of hydrogen ions and thus offering a high resistance to the flow of current.

deprotonated Having had a proton removed. For example, a deprotonated water molecule is the hydroxyl ion OH^-.

depth of field The distance on either side of the point of best focus of an optical instrument within which objects are acceptably well focused.

depth of focus The range of distance between a lens and a focal plane in which the image remains acceptably sharp.

desorption The opposite of adsorption.

detector count rate In an instrument that detects ionizing radiation, the count rate indicates the strength of the ionizing radiation.

deuterium The isotope of hydrogen that has a neutron as well as a proton in the nucleus: heavy hydrogen, or hydrogen-2, with a mass number of 2. It occurs as deuterium oxide (heavy water), from which it can be obtained by fractional distillation or electrolysis. Deuterium, having the same single electron as hydrogen, behaves chemically in a very similar way, but its physical properties are slightly different.

deuteron The nucleus of a deuterium atom.

deviation of light ray The change of path of a light ray passing from one medium to another.

device driver A software program that controls a device such as a printer, a disk drive, a keyboard, a mouse, or a modem. Devices will operate correctly only if the commands they understand are issued. Device drivers translate computer instructions into these correct commands.

Dewar flask A flask designed to prevent its contents losing or gaining heat by conduction (the walls of the flask contain a vacuum), radiation (the walls are silvered), or convection (the contents are isolated so there can be no convection currents).

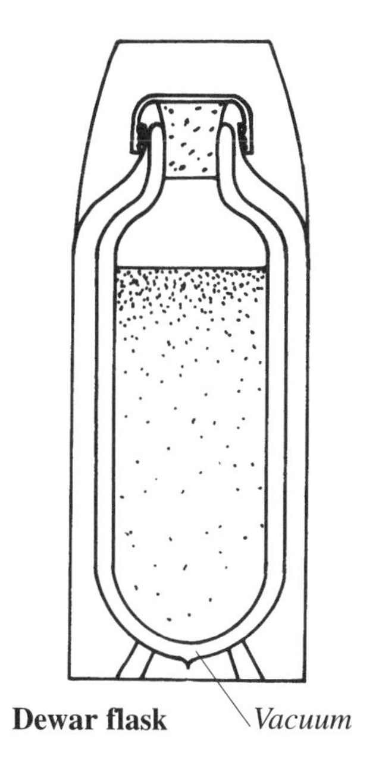

Dewar flask

diamagnetism The domains within which diamagnetic materials align themselves at right angles to the lines of force in a magnetic material.

diameter A straight line that connects two points on the perimeter of a circle to its center.

diaphragm Any thin dividing membrane. In photography and optics, a membrane with a central hole of variable or adjustable size that is used to reduce the amount of light passing through.

dielectric A substance having very low electrical conductivity, used as an insulator. In a capacitor, the air or insulating material between plates is the dielectric. The application of an electric field to a dielectric causes a displacement of electrons relative to the atomic nuclei, but not a flow of electrons. The result is a dipole with an electric moment in the direction of the field. Dielectrics situated between the plates of a capacitor increase the capacitance by a factor known as the dielectric constant or relative permittivity. The dielectric strength is the volts per millimeter that can be applied to the material without causing it to break down.

dielectric heating A method of heating a dielectric material by placing it between metal plates and applying a high-frequency alternating current to them so that an oscillating electric field is formed through the dielectric, causing rapid movement of the electrons relative to the atomic nuclei.

differential equation An equation that contains a derivative of any quantity y with respect to another quantity x, as well as the variables x and y. There are many different kinds of differential equation, each solved in its own way. Differential equations usually have to be solved using numerical methods because they cannot be solved using exact (analytical) methods.

diffraction Spreading or bending of waves (electromagnetic, sound, water) that pass the edge of an obstacle or pass through a small opening. The extent of bending is large if the obstacle or opening is about the same size as the wavelength of the wave. Light diffracted by a narrow slit produces a set of interference fringes.

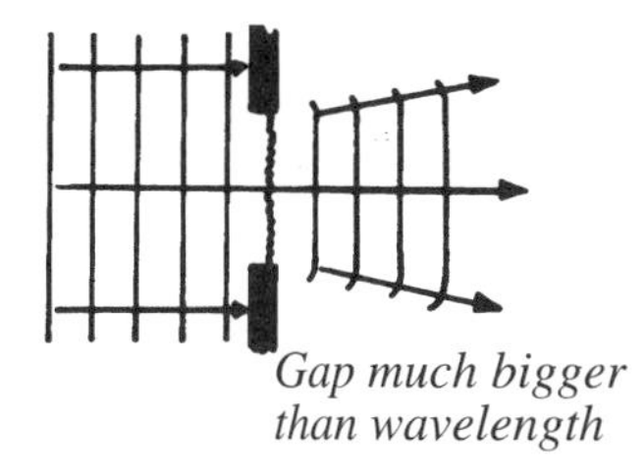

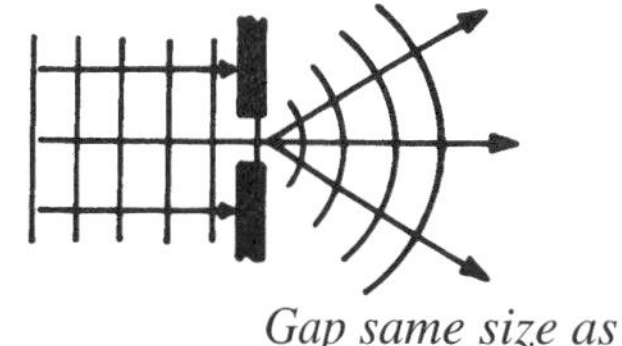

Diffraction

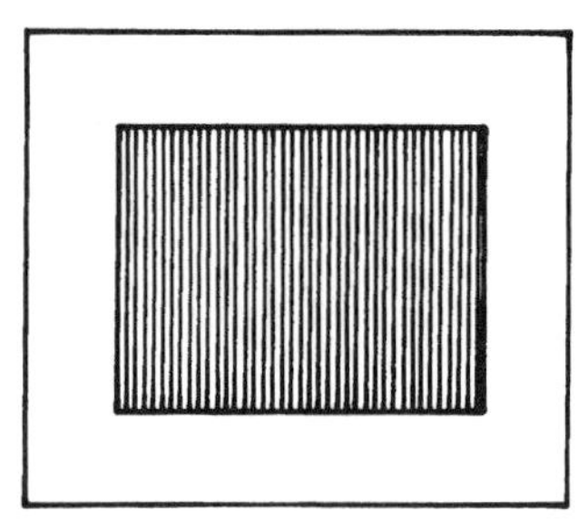

Diffraction grating

diffraction grating A piece of transparent film or glass printed or engraved with many thousands of closely spaced opaque parallel lines. Different wavelengths of light diffract differently, forming spectra and multiple images.

diffusion The process of rapid random movement of the particles of a liquid or gas that eventually form a uniform mixture.

digital (1) Pertaining to the method of representing information of any kind as a sequence of discrete numbers, rather than by a continuous varying quantity (analog).
(2) Of any system based on discontinuous events. Digital representations are only approximations of analog events, but the approximation can be made ever closer, by increasing the number of samples taken in a given time, until the difference is imperceptible. The information on a music compact disk is purely digital, but because the music was sampled some 40,000 times a second, the loss of detail is negligible. After digital to analog conversion, the original sound is recreated with remarkable fidelity, and the discontinuity of such short intervals cannot be appreciated. Digital computers handle only digital information, and that in binary form. During transmission, digitalized information is inherently much less prone to interference and distortion than analog information. *Compare* analog.

digital code A way of displaying information as numerical values. For electronics in which inputs and outputs can be only 0 or 1, it is a way of displaying data and instructions with a combination of these digits.

dihedral (1) The angle formed when two planes, such as those on the surface of a crystal, intersect.
(2) In aeronautics, the slight upward tilt of the wings of an aircraft that reduces the tendency to roll.

diode A component of an electrical circuit that allows current to pass in only one direction. Most diodes are semiconductors and are used for signal demodulation, rectification of alternating current, voltage regulation (Zener diodes), and other purposes.

diopter A measure of lens power. Lens power can also be expressed as the focal length – the distance from the lens to the point at which parallel rays are brought to a focus. But the focal length becomes inconvenient as a measure of lens power when lenses are combined. Two lenses of focal length 10 cm have a combined focal length of

5 cm. A lens of 20 cm in combination with a lens of 10 cm produces a focal length of just below 7 cm. To calculate the focal length of the combination, it is necessary to take the reciprocal of each focal length (1 divided by the length), add them, and then take the reciprocal of the result. To get around this difficulty, thin lenses, such as are used in eyeglasses, are graded, not by focal length, but by the reciprocal of the focal length. This is called the diopter. A lens of one meter focal length has a power of one diopter. A lens of 50 cm focal length has a power of two diopters, and one of 20 cm focal length has a power of five diopters. The dioptric power is easily obtained by dividing 100 by the focal length in centimeters.

dipole (1) Two point charges, or poles, equal in magnitude but opposite in sign, separated by a small distance.
(2) A molecule, such as the water molecule. This is a dipole because the electrons in the bonds move nearer the more electronegative oxygen atom, giving the other end of the molecule a more positive charge.
(3) In radio transmission and reception, a dipole is a form of antenna routinely used for high frequencies and dimensioned to correspond to the desired wavelength. It consists of two short rods arranged in a straight line, with the inner ends close together but insulated from each other. The connections are made to these inner ends.

direct current (DC) An electric current that flows in only one direction.

directed line segment A way of representing a vector geometrically.

direct proportion The linear relationship wherein the ratio between two properties in direct proportion is constant.

disassembler A program that attempts to convert computer machine code to assembly language – a sometime disreputable process known a reverse engineering.

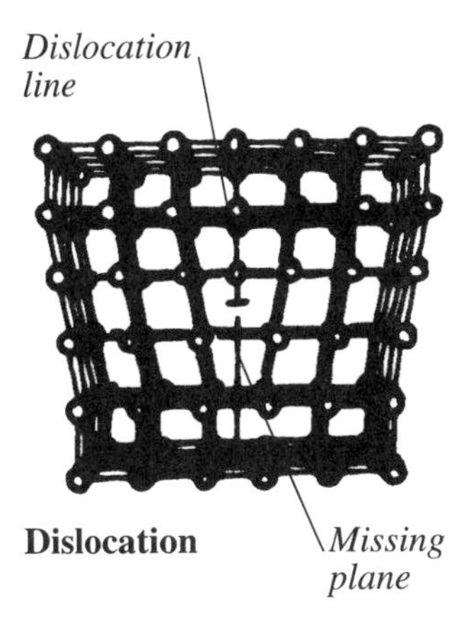

Dislocation

dislocation An area of a crystal lattice in which a plane of atoms or ions is missing. As a result, the bonding is weaker, which allows the crystal to be more malleable (this is of special importance in metals).

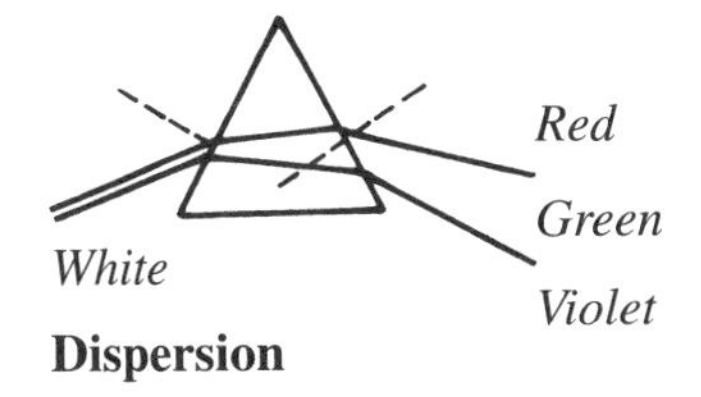

Dispersion

dispersion (1) The separation of a beam of mixed frequencies, for example, the separation of a beam of white light by a prism to form a spectrum. Waves of different frequency traveling in a dispersing medium are separated because they travel at different speeds. The degree of bending depends on the wavelength of the light; shorter wavelengths of light are bent more than red. Dispersive power is the ability of a medium to disperse rays.

(2) A colloidal dispersion is an even distribution of small particles in a fluid.

displacement (1) In a chemical reaction, one element (or group of elements) taking the place of a less reactive one.
(2) The amount of fluid displaced by a body floating on or submerged in a fluid (*see* Archimedes' principle).
(3) In wave motion, a disturbance from zero at any given time. The maximum displacement is the wave's amplitude.

displacement vector The vector describing an object's change of position.

dissociation The breakdown of molecules to smaller molecules, atoms, or ions. It can be electrolytic or thermal and can be reversible. The dissociation constant at a given temperature gives the relationship between the degree of dissociation and the concentration of the solution.

dissolve To add a solute to a solvent to form a uniform solution.

Distillation flask
Thermometer
Water to waste
Water from tap
Receiver
Distiller

Distillation

distillation (1) A process in which a solution (or a mixture of liquids whose boiling points are widely differing) is heated to a particular temperature to produce a vapor of one of the components. This vapor is condensed, forming a pure liquid that has a single boiling point.
(2) In statistics, a set of possible values of points in a sample space.

distribution A term used in Boolean algebra that describes one operator's ability to be distributed over the other.

diverging lens *See* concave.

diverging mirror *See* convex.

diverging rays Narrow beams of radiation that move away from each other.

domain A small naturally magnetized area in a magnetic material. Such a magnetic material contains large numbers of such domains, usually arranged randomly. When magnetized, all domains in the material point in one direction, leaving free domains at each end, that form the poles of the magnet.

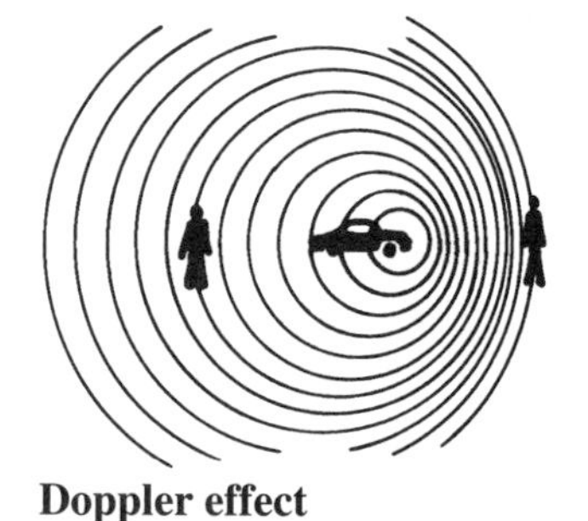

Doppler effect

Doppler effect The apparent change in frequency of wave motion when there is relative motion between the source of a sound and an observer. This effect is demonstrated by a police siren. It has a higher pitch when approaching the observer than when moving away. A red shift in visible radiation has the same cause.

dose equivalent (of ionizing radiation) The effect of ionizing radiation on tissue. Dose equivalent allows for the different ionizing abilities of different types of radiation.

downloading The receipt of data by a remote computer from a central or controlling machine, by way of a local connecting network, modem, or other communication channel.

drag coefficient A measure of the resistance to the movement of a body through a fluid medium, such as air. Air resistance to movement varies with the square of the speed of the body. In a motor vehicle, the power needed to overcome drag increases with the cube of the speed. Drag increases with the frontal area of the vehicle and with its shape. Streamlining and reduction of the frontal area can reduce drag coefficient considerably.

drop-out The loss of data because of local defects on the surface of a magnetic recording medium, such as a floppy or hard disk.

dry ice Solid carbon dioxide. Carbon dioxide solidifies at –78.5°C. Dry ice turns (sublimes) directly into a gas if it is heated above this temperature.

ductility The ability of a metal to retain its strength even when substantially deformed or drawn into wires.

duplex operation The transmission of data in both directions at the same time, as in an ordinary telephone conversation. Half-duplex allows transmission in only one direction at a time, as in earlier walky-talky radio operation.

dynamic equilibrium A balanced state of a changing system. A reversible chemical reaction may reach a state of dynamic equilibrium when the rate of the forward reaction is equal to the rate of the backward reaction.

dynamic random access memory (DRAM) A semiconductor memory in which bits are stored as charges on the small capacitance in the gate of a metal-oxide silicon transistor. These charges quickly leak away and must be refreshed at short intervals. Such memories are called dynamic random access memories (DRAMs). DRAMs can store many millions of binary digits (bits).

dynamics The branch of mechanics concerned with the motion of bodies and the forces on them that cause motion or change in motion.

dynamo A machine that converts mechanical energy to electrical. Dynamos can produce direct or alternating current, depending on their design. The large generators in power stations are alternators.

dynamometer An instrument, such as a spring balance, used to measure a force.

earth (*electrical*) The electrical connection by a conductor between an electrical appliance and the Earth (assumed to be at zero potential). It protects the user from electric shock if a fault in the appliance allows electricity to flow through the casing of the appliance.

Earth's magnetic field

Earth's magnetic field The field is similar in shape to that of a bar magnet, with its poles close to geographic North and South Poles. The field extends many thousands of miles above the surface. It may be caused by convection currents in the Earth's core.

echo The reflection of a wave of any kind by a surface.

Edison effect The electrical conduction between an incandescent filament and a separate cold electrode that is made positive with respect to the filament. This effect was the basis of the thermionic diode, from which was developed the triode and, with it, the beginnings of electronics.

eddy current The electric current that is induced in a conductor cut by a changing magnetic field. If the resistance of the conductor is low, as in the case of a block of metal, the eddy currents can be very large and can cause heating and considerable energy loss. For this reason, transformer cores are made of stacks of thin sheets of metal, each sheet being insulated.

effervescence The production of bubbles that rise to the surface in a liquid.

efficiency The ratio of the energy output of a machine to the energy input. Efficiency usually varies with the load and is a measure of the extent to which avoidable losses have been avoided. Thermal efficiency is the ratio of the work done to the heat supplied by the fuel.

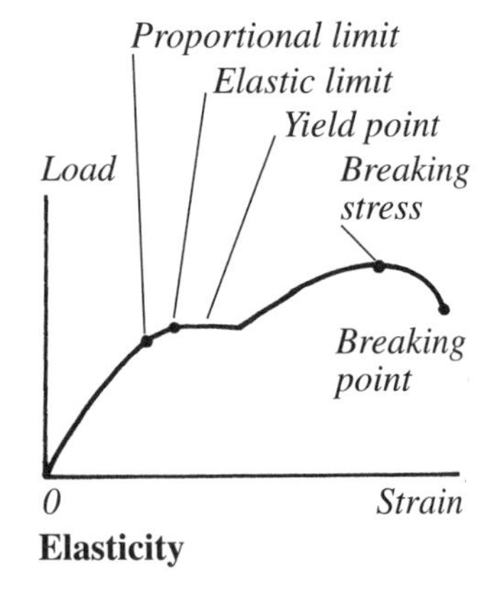

Elasticity

EISA Acronym for *e*xtended *i*ndustry *s*tandard *a*rchitecture, a microcomputer bus structure for machines using Intel 32-bit processors, such as the MS-DOS personal computers.

elastic collision A collision in which no loss of kinetic energy occurs. Elastic collisions do not occur between particles or bodies larger than atoms.

elasticity The property of a body that allows it to return to its original shape and size after distortion by a force. Different moduli of elasticity are observed, depending on the type of force applied to a body. The bulk modulus applies when the body is compressed or dilated. Young's modulus applies when the body is under tensional stress. The modulus of rigidity, or shear modulus, applies when the body has been pulled out of shape by a shearing force.

elastic limit If a force greater than the elastic limit is applied to a body, the body does not return to its original shape. If the force applied is less than the elastic limit, the body returns to its original shape when the force is removed.

electret A body with a permanent electric charge, analogous to the permanent magnetic field of a permanent magnet. Electrets can be used in microphones.

electrical energy The energy acquired from moving electric charges.

electrical force The force exerted on an electric charge in an electric field.

electrical potential difference *See* potential difference.

electric capacitance *See* capacitance.

electric charge *See* charge.

electric conductance *See* conductance.

electric conductivity *See* conductivity.

electric current *See* current.

electric current density The amount of current flowing per unit cross-sectional area of a conductor.

electric dipole moment *See* dipole.

electric field An area in which forces are exerted on electric charges.

Unequal field and opposition charges

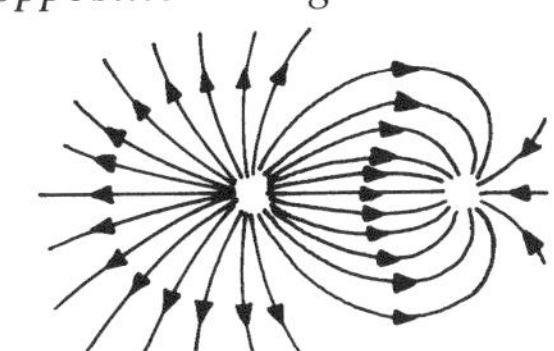

Electric field

electric field strength or **intensity** An electric field is produced around a charged object. The higher the charge, the more intense the field strength. Electric fields can be described as having lines of electric force or field lines.

electricity The term used to describe the effects of electric charges. Moving charges cause electric current, stationary charges cause static electricity.

electric motor A device that converts electrical energy to mechanical energy. It makes use of the movement of a current-carrying coil (armature), which can turn in a magnetic field.

electric potential *See* potential.

electric potential difference *See* potential difference.

electric resistance A measure of the extent to which an element of a circuit opposes the passage of an electric current. It depends on the nature of the material. *See* Ohm's law.

electric wind The stream of ionized and unionized gaseous molecules that flows away from a highly charged electrical conductor. Molecules of gas are ionized by colliding around the conductor. Ions of the same charge as the conductor are repelled and move away as electric wind, taking uncharged gaseous molecules with them.

electrification Charging with electricity.

electrochemical series (electromagnetic series, displacement series) Metallic elements arranged in order of increasing electrode potential (or readiness to release electrons and form cations), the highest having large negative values. The more reactive a metal is (the higher in the table), the greater the force with which electrons move. The series can be used to predict reactions by comparing reactivities.

electrode A conductor that allows current to flow through an electrolyte, gas, vacuum, dielectric, or semiconductor.

electrode potential The potential difference between an element and an electrolyte that contains its ions.

electrolysis The process by which an electrolyte is decomposed when a direct current is passed through it between electrodes. Positive cations move to the cathode to gain electrons; negative anions move to the anode to lose electrons. Substances are either deposited or liberated at the electrodes, depending on the nature of the electrodes and electrolyte.

Negative ions move to anode
+ve
– ve
Positive ions move to cathode
Solution of $CuCl_2$

Electrolysis

electrolyte A substance that forms ions when molten or dissolved in a solvent and that carries an electric current during electrolysis. Strong electrolytes contain many ions.

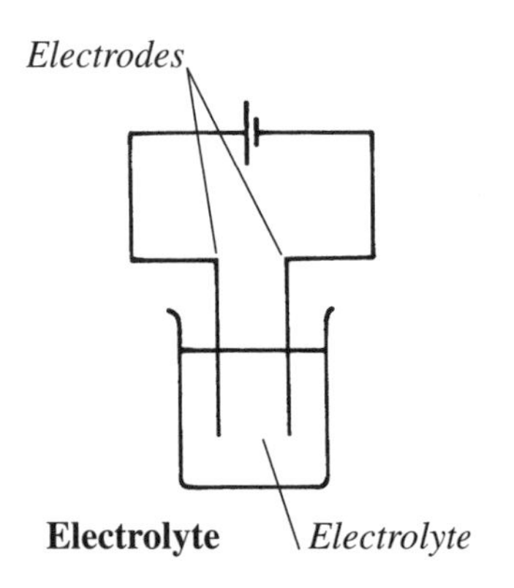

Electrolyte

electrolytic capacitor A type of capacitor that is small in size but gives large capacitance values. The capacitance of a capacitor is inversely proportional to the separation of the plates, so if these can be brought very close together, capacitance is increased. In an electrolytic capacitor, a very thin film of oxide is deposited on one plate and this is immersed in a liquid that acts as the other plate. The metal plate is also deeply etched so as greatly to increase its surface area. Electrolytic capacitors must be connected correctly with respect to electrical polarity.

electromagnet A magnet produced by the effect of a current flowing in a coil surrounding a soft-iron core.

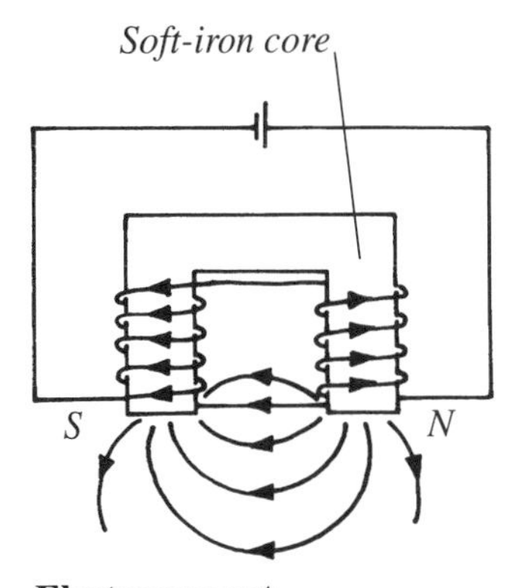

Electromagnet

electromagnetic induction The production across a conductor of an induced e.m.f. that moves relative to a magnetic field. One way in which this is done is to move a magnet toward and away from a coil of wire (induction coil). E.m.f. is induced when the coil and magnet are moving relative to each other. *See* Lenz's law.

electromagnetic moment This gives the magnitude of the magnetic strength of a magnet or current-carrying coil. It is defined as the torque produced when a magnetic coil is perpendicular to unit magnetic flux density.

electromagnetic radiation The energy that results from the acceleration of electric charges. This energy is radiated through space (vacuum) at the speed of light, in the form of oscillating electric and magnetic fields at right angles to each other and to the direction of propagation. Through media the speed is slower. Electromagnetic radiation can also be regarded as a stream of photons traveling at the speed of light, each with an energy equal to Planck's constant multiplied by the speed of light, divided by the wavelength of the associated wave. Electromagnetic waves travel at a speed of 3.00×108 m/s^{-1} and do not need a medium through which to travel. Like all wave motions, they show reflection, refraction, interference, and diffraction. The properties of the radiation depend on its wavelength.

electromagnetic spectrum The range of frequencies of electromagnetic radiation. Gamma radiation has wavelengths below 10^{-12} m; X rays, between 10^{-12} and 10^{-10} m; ultraviolet, between 10^{-10} and 10^{-7} m; visible light, between 10^{-7} and 10^{-6} m; infrared, between 10^{-6} and 10^{-3} m; microwaves, between 10^{-3} and 10 m; radio waves, between 10 and 10^{6} m.

electromagnetism The magnetic force produced by electricity, and the electric effects produced by magnetic fields.

electromotive force (e.m.f.) The power of an electric cell to push electrons around a circuit.

electron One of the three basic subatomic particles. It is very light (its mass is 9.109×10^{-31} kg) and orbits around the nucleus of an atom. It has a negative charge, and in neutral atoms, the number of electrons is equal to the number of protons in the nucleus.

electron beam lithography A method of etching very fine lines using a powerful beam of focused electrons; used to prepare the masks for the production of electronic microchips.

electron diffraction Diffraction of a beam of electrons by atoms or molecules. Louis de Broglie predicted that electrons should behave like waves and it should therefore be possible to diffract them in the same way as light. This was demonstrated in 1925 by C. J. Davisson. When he bombarded nickel with low-energy electrons, the surface of the nickel, which had a large crystalline structure, behaved like a diffraction grating. From this Davisson calculated the wavelength of electrons.

electronegativity A measure of the ease with which an atom can attract electrons. Group 7 of the periodic table contains electronegative elements, fluorine being the most electronegative.

electronic configuration Atoms of each element contain a certain number of electrons, which are found in shells around its nucleus. Each shell can contain a certain number of electrons. The first (the closest shell to the nucleus) can contain two, the second shell can contain eight, and the third can also contain eight. The degree to which these shells are full affects the properties of the element. Elements that have a full outer shell tend to be unreactive. (All the noble gases, for example, have a full outer shell.)

electronic mail or **email** The multi-way transmission of personal and business correspondence via digital communication channels. A letter is written, using a word processor or simple editor, and, with communications software run on the personal computer, is directed to one or many other correspondents, situated in the same office or halfway around the world. To each contributor the Internet service provider offers a personal mailbox where the letter will remain until the addressee checks his or her mail, when it is retrieved. Printing and mailing are unnecessary and copies can be kept on disk for later perusal by the sender and all recipients.

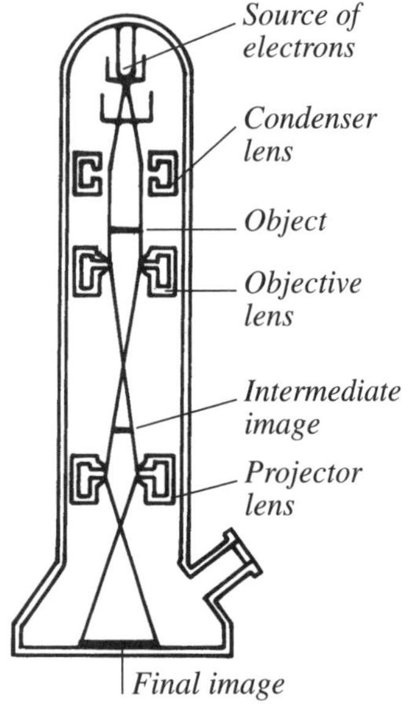

Electron microscope

electron microscope A microscope that uses a beam of high-energy electrons rather than light. The beam is focused magnetically or electrostatically. As the wavelength of electrons is less than that of light, the electron microscope can produce more detailed images than optical microscopes.

electron shell Electrons appear in a series of cloudlike shells around the nucleus. The shell is not full of electrons, but is an area in which there is a high probability of finding one or more electrons.

electronvolt (eV) A unit of energy equal to the work done in moving an electron through a potential difference of one volt. The electronvolt is used as a measure of the energy of particles. When an electron passes through a potential rise of one volt, the eV is the quantity of energy it gains. 1 eV $=1.602 \times 10^{-19}$ joules.

electrophile Particles (ions or molecules) that can accept a pair of electrons to form a new bond because their outer electron shells are short two electrons required to form a stable (full) outer shell.

electrophoresis The process by which charged particles (ions) in a solution are separated by the application of an electric current. Many substances ionize naturally when dissolved to give positive and negative particles. If an electric current is applied to the solution, the positive ions are attracted to the negative electrode and the negative ions to the positive electrode. In electrophoresis, the solution of the ions is spread out on a surface, such as a filter paper or a gel, and the electric current is applied across it. The movement of the ions occurs on the surface, and the extent of the movement is dependent on the molecular weight and the charge of the ions. Ions of low weight move more quickly than those of high weight, so separation occurs. After a time, the surface can be stained so that the characteristic patterns of separation can be recognized. In this way, mixtures of substances can be separated. Electrophoresis is used, for instance, to separate mixtures of proteins or different fragments of DNA.

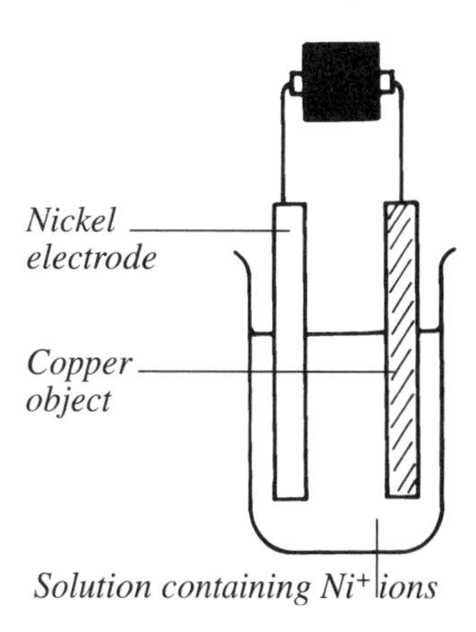

Electroplating

electroplating Electrolytic coating of a metal with a less reactive one. The metal to be plated is used as the cathode in an electrolyte containing ions of the metal that is used for the plating. These ions are deposited firmly on the surface of the cathode.

electropositivity A measure of the ease with which an atom loses electrons. Elements from Group 1 of the periodic table are all very electropositive.

electroscope A device used to detect if an object is charged and, if so, whether the charge is positive or negative.

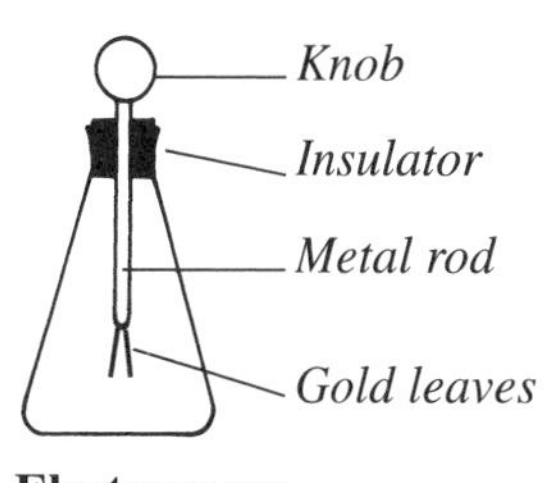

Electroscope

electrostatic precipitator Used to clean gas emissions of suspended solid particles. These particles are given an electric charge, which then allows them to be attracted to charged plates.

electrostatics The study of the effects of stationary electric charges.

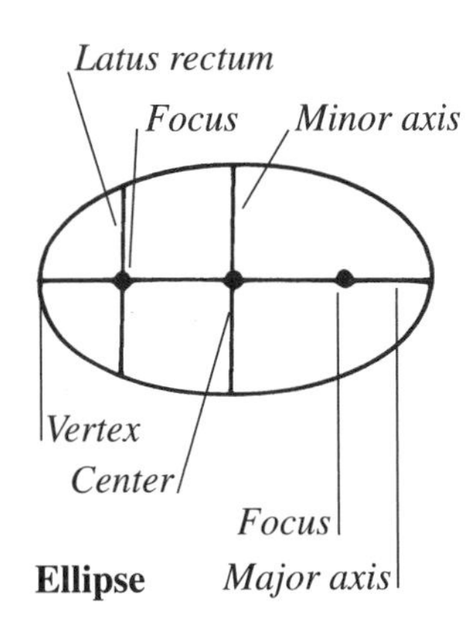

Ellipse

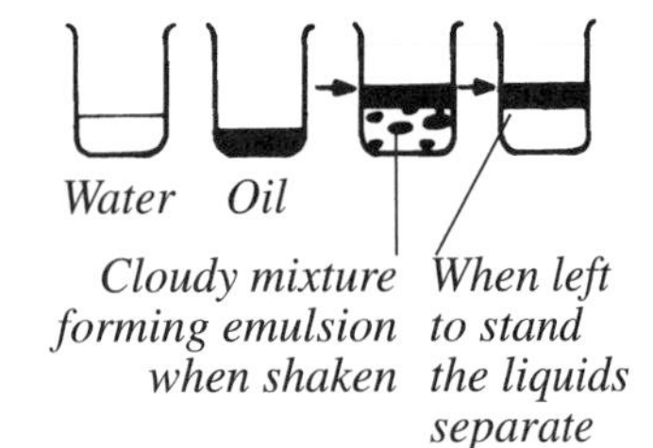

Emulsion

electrostriction The change in the dimensions of a body when it is placed in an electric field.

element A pure substance that cannot be split chemically into simpler substances. Only one type of atom is found within a pure element. Over 100 elements have been discovered, 91 of which occur naturally.

elementary (electron) charge When studying charges on subatomic particles, the charge on the electron (elementary charge) is used in simple multiples (positive or negative) to describe the charge on the particle.

ellipse A flattened circle that can be formed by taking a slice through a cone, so long as that slice is not parallel to the base of the cone.

email *See* electronic mail.

e.m.f. Abbreviation of electromotive force.

emission spectrum Produced by matter emitting electromagnetic radiation. Line spectra are observed as a pattern of radiation emitted by pure elements that are heated or electrically disturbed. Each element has a unique spectrum.

empirical formula This gives the simplest ratio of atoms present in a compound (the molecular formula may be a multiple of its empirical formula).

emulator An item of hardware or, more often, software, by the use of which a computer of one kind can be made to behave as if it were of another kind so that programs intended for the latter can be run.

emulsion A colloidal dispersion of small droplets of one liquid within another, such as oil in water or water in oil.

endothermic change A chemical reaction that absorbs heat from the surroundings.

energy The ability of a system to do work. There are two fundamental forms of energy: potential (due to position) and kinetic (due to motion). Other forms of energy – chemical, electrical, heat, light, mechanical, nuclear, sound – can be described in terms of the fundamental forms and are largely interchangeable.

energy bands Bands of allowed energies, each representing many quantum states, that electrons can exist in, in a crystalline solid. Between the energy bands are forbidden bands. The valence band, that with the highest energy, is formed from the outer electrons of the atoms – those participating in chemical bonds. The electrical properties of solids are determined by their band structure. Electrical conduction can occur only if there is an unfilled band – the conduction band – and this must be close to, or overlap, the valence band. In insulators these bands are separated by a wide forbidden band that electrons do not have enough energy to cross. Semiconductors depend on energy bands for their function.

energy cycle If there are two ways in which a chemical reaction can take place, the same quantity of energy is used for each way, regardless of the number of steps involved. If the two reactions are put together diagrammatically, the clockwise changes are equal to the counterclockwise ones.

enthalpy A measure of the stored heat energy of a substance. During a chemical reaction, change in enthalpy can be measured. If energy is released, the reaction is exothermic; if energy is absorbed, the reaction is endothermic.

entropy A way of describing the disorder in a system. The more disordered the system, the higher its entropy.

epitaxy A way of forming a thin layer of a material, such as gallium arsenide, on a base of another material so that semiconductors can be formed. Gallium arsenide, for instance, can be deposited by firing ions of the compound onto the base.

EPROM Acronym for *e*rasable *p*rogramable *r*ead-*o*nly *m*emory, a nonvolatile ROM whose stored material can be changed by the user.

equilibrium (1) A state of stability where forces acting on a body cancel each other out.
(2) The state of a reversible chemical reaction in which the forward and backward reactions take place at the same rate (i.e., equilibrium is reached when there is no apparent change in the amounts of reactants and products with time).
(3) Thermal equilibrium is the state of a body in which no net heat flow occurs between it and its surroundings.

erase To delete a computer file from the storage medium – usually a magnetic tape or disk – on which it has previously been saved. In the MS-DOS system, deletion does not remove the magnetic record but merely frees the part of the disk so that it can be overwritten by subsequent data. At any time before this is done, the erased file can be retrieved by an undelete program. If another file is saved to the same area, however, the data is irretrievably lost.

escape speed The speed that a body must attain in order to escape from the gravitational attraction of another body. The escape velocity from Earth is 11,200 meters per second.

eureka can A metal can that has an overflow tube. It is used to measure the volume of irregular solids. If the can is filled with water to the base of the overflow tube, when the object is lowered into the can, the volume of the displaced water is equal to the volume of the object.

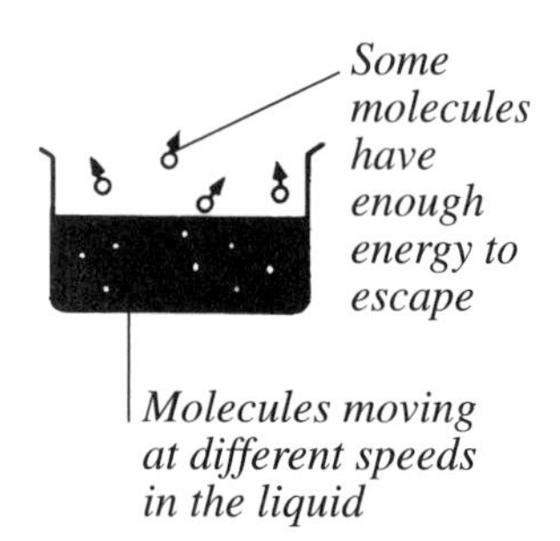

Evaporation

evaporation The process in which a liquid changes state to vapor. It can occur at any temperature up to the boiling point of the liquid. An evaporator is a system in which evaporation can take place.

executable file A file that a computer can run as a functioning program and which has been compiled or assembled from the original source code. Such a file will usually have the extension .EXE.

exothermic change A chemical reaction that releases heat to the surroundings.

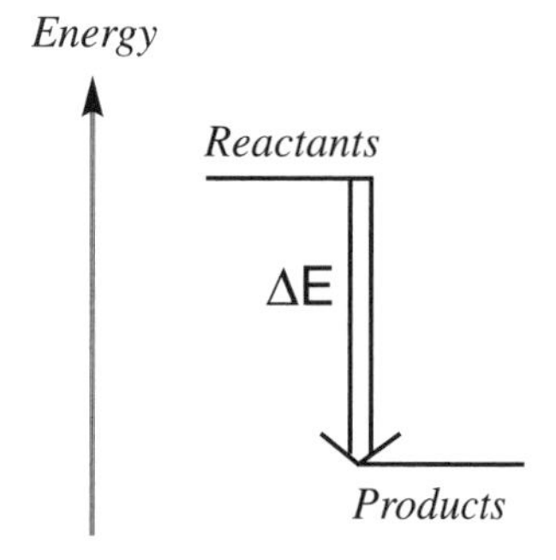

Exothermic change

expansion A physical process in which there is an increase of an object's length, area, or volume.

expansion card A circuit board card that can be plugged into a personal computer to increase its capabilities in one of many directions. Expansion cards may confer additional memory, extra disk controllers, modems, fax facilities, CD-ROM facilities, and so on.

explosimeter An application of Wheatstone bridge used to measure the strength of an explosion.

explosion A rapid increase in pressure caused by the large volume of gas produced as a result of a violent chemical or nuclear reaction.

exponential A function that varies with the power of another quantity. In the equation $y = a^x$, y varies exponentially with x. In exponential growth of a group, the rate of growth is related to the number of individuals in the group, and is slow when the number is small and increasingly rapid as the number increases.

exponential decay If a property, such as the radioactivity emitted by a radioactive substance, decays by the same ratio in equal time intervals, it is said to undergo exponential decay.

extraction (1) The process of removing a substance from a solid or liquid mixture.
(2) The process of removing a metal from its ore.

facsimile (fax) transmission Electronic transmission of documents of any kind, including photographs, by a scanning system that converts the image into an analog or digital signal which is passed to the receiving end, where the image is reconstituted on paper. Telephone lines are commonly used for fax transmission.

Faraday constant (F) The amount of electricity equal to one mole of electrons. Also, the quantity needed to liberate one mole of a monovalent ion during electrolysis. It is equal to $9.648\ 670 \times 10^4$ coulombs per mole.

Fahrenheit scale The temperature scale in which the melting point of ice is 32° and the boiling point of water is 212°. To convert Fahrenheit to Celsius, subtract 32 and multiply by 0.555 or 5/9.

fallout The deposition of radioactive particles produced by a nuclear explosion and carried by prevailing winds in the atmosphere, often a long distance from the site of origin. The yield from high explosions may fall anywhere on the surface of the Earth and may continue for years. Fallout of isotopes with a long half-life, such as strontium-90 and iodine-131, is the most significant as both will enter the food chain and have effects for many years.

feedback (*electronics*) The return to an amplifier as an input of a signal obtained from the output of the amplifier. If this returning signal is in phase, the gain is increased; if it is in antiphase, the feedback is negative, producing lower gain but reduced noise and distortion from the amplifier.

Fermat's principle The important proposition by the French mathematician Pierre Fermat that the path taken by light between two points is always that which requires the least time. The laws of optics are based on this principle.

fermions Subatomic particles whose spin can be expressed in half integers and obey Fermi-Dirac statistics. Protons, neutrons, and electrons are fermions. Particles whose spins can be expressed as whole numbers are called bosons.

ferrites A group of magnetic ceramic compounds of the oxides of metals, such as cobalt, manganese, or zinc, linked to iron oxides. Ferrites are ceramic materials but not electrical conductors and avoid large eddy currents. They are useful as magnetic cores in high-frequency circuits, providing high inductance with low resistance and low losses.

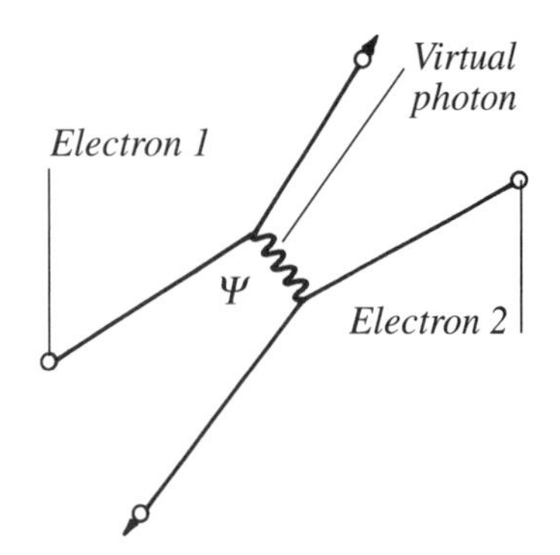

Feynman diagram

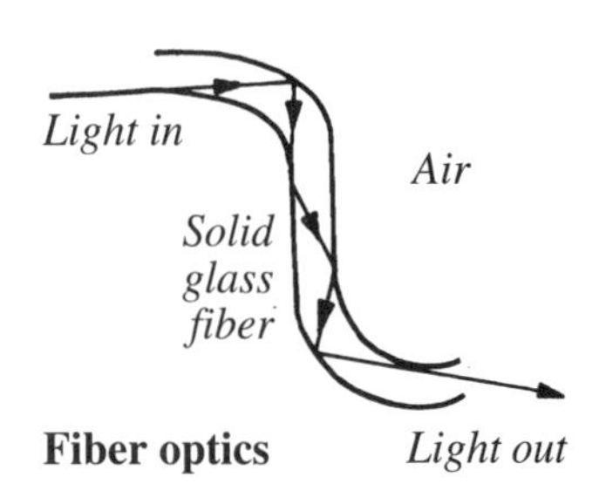

Fiber optics

ferromagnetic A material, such as iron, steel, nickel, or cobalt, that can be made into a strong magnet. Atoms of these elements contain unpaired electrons; this gives the atom a net magnetic moment. If the element is placed in an external magnetic field, its atoms will align with that field, responding strongly to it.

Feynman diagram A graphical representation in terms of quantum mechanics of the space-time events – the exchange of photons – involved in the interaction of subatomic particles such as the collision of a moving electron with a proton. Feynman diagrams are used in quantum electrodynamics to study the way electromagnetic radiation interacts with charged matter.

fiber optics A branch of optics concerned with the transmission of light along optical fibers – fine, flexible rods of glass or other transparent materials. Fiber optics has become increasingly important in communications and in situations requiring direct visual inspection of inaccessible cavities as in medicine and surgery. Optical fibers are used to guide light – both for illumination and for viewing – around complex bends or sharp corners by making use of the principle of total internal reflection. Fibers with a diameter of as little as 0.02 mm are arranged in tight bundles and, so long as the fibers remain in registration at both ends of the bundle, they can be bent and maneuvred into otherwise inaccessible places.

field ion microscope A type of microscope used to examine the atomic structure of surfaces by observing the desorption of ions from the surface in a high electric field.

field lens A lens placed in or near the plane of an image in an optical system. It ensures that all the light from the image proceeds to the subsequent lenses in the system, thus ensuring uniform illumination over the field of view.

filament A thin, high-resistance wire through which electric current passes. This causes the filament to heat; in an electric light bulb the filament becomes white hot.

film badge A method of monitoring the risks of excessive radiation dosage in people working in environments in which exposure to radiation is possible. The film badge is a small light-tight container for a piece of photographic film that becomes fogged if exposed to radiation. All personnel at risk carry film badges, usually pinned to the clothing, and the films are regularly developed and replaced. A fogged film indicates that an accident has occurred and this can be investigated.

filter A device containing a porous material, such as paper or sand, that removes suspended solid particles from a fluid.

filtrate Clear liquid that has passed through a filter.

first law of thermodynamics *See* conservation of energy, law of.

fission A process (spontaneous or induced) during which a heavy atomic nucleus disintegrates into two lighter atoms that together have less mass than the total initial material. This lost mass is converted to energy – the amount is given by Einstein's equation $E = mc^2$.

Fizeau's rotating wheel A device used to measure the speed of light, invented by Armand Fizeau in 1849.

flame test A test to detect the presence of metal ions in a compound. A wire is dipped into a sample of the compound and then placed in a flame. The color of the flame identifies the ions present in the sample because different elements emit characteristic wavelengths of light when they are heated.

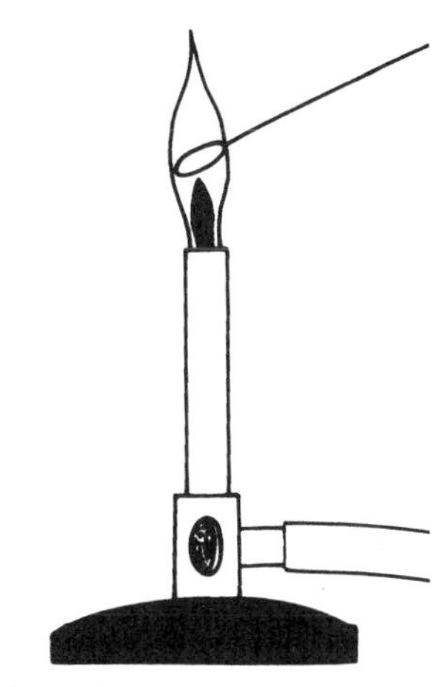

Flame test

Fleming's right-hand rule A mnemonic for the directions of the field, current, and force in an electrical generator. The forefinger points forwards, the thumb upwards, and the second finger to the left, all three being at right angles to each other. If the second finger shows the direction of the current, the forefinger will show the direction of the magnetic field, and the thumb will show the direction of the force. The rule was proposed by the English electrical engineer Sir John Ambrose Fleming. The analogous left-hand rule applies to electric motors.

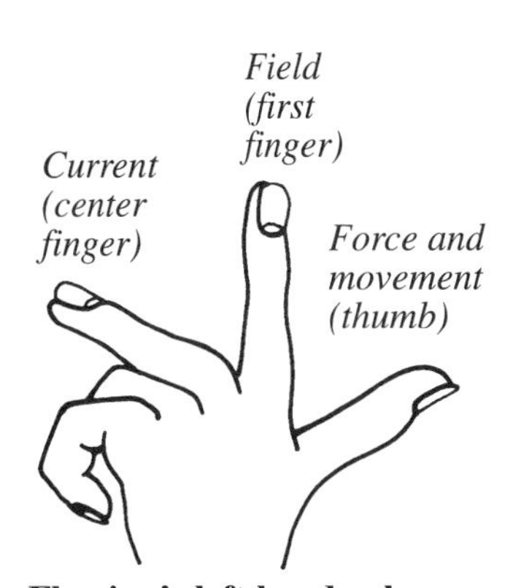

Fleming's left-hand rule

flip-flop An electrical device whose output changes between zero and one each time an input pulse is applied.

floating An object is said to be floating if it rests on the surface of a fluid without sinking. The upthrust is greater than or equal to the mass of the object.

floppy disk A thin disk of flexible plastic coated with a magnetic material that rotates at high speed within a suitable protective container when the whole is slotted into a personal computer disk drive. An electromagnetic read-write head bears on the surface of the disk to "play back" previously recorded binary data or to save a copy of data produced in, and temporarily or permanently stored in, the machine.

Floppy disk

fluid A substance that can flow because its particles are not fixed in position. Liquids and gases are fluids.

fluorescence The emission of radiation from certain substances that have been irradiated by light or other radiations. The light emitted is at a lower frequency than the original radiation. Fluorescent substances are those that can emit fluorescence. A fluorescent tube generates light by fluorescence. The inside of the tube is coated with a fluorescent substance and the ultraviolet radiation formed inside the tube is converted to visible light by this fluorescent coating.

flux (*magnetic or electrical*) The strength of a field (magnetic or electrical) in a given area is the product of the area and the part of the field strength that is at right angles to the area.

flux change If there is a movement in a coil of wire in a magnetic field, e.m.f. is induced in the wire. The size of the induced e.m.f. in this coil of wire is proportional to the rate of change of flux.

focal length The distance from the center of a lens to its focal point.

focus or **focal point** *See* principal focus.

Force

force (1) Any influence that causes a resting body to move or that changes the momentum of a moving body. The magnitude of the force is equal to the product of the mass of the body and its acceleration. (2) Any influence that produces an elastic strain in a body or system or that sustains weight.

formatting The initialization of a computer disk by erasing the addresses of all prior material, recording on it a standard pattern of magnetic sectors to receive data, the checking of these sectors, and the noting of those that are unreliable so that they will not be used. Files cannot be saved to computer disks until they have been formatted; this is done by issuing the format command that runs a formatting program.

fossil fuel A fuel such as coal, petroleum, or natural gas that was formed by the decomposition of living organisms in prehistoric times.

fractal A curve or area produced by repetitive reproduction of the same polygonal or polyhedral shape in fixed subdivided dimensions.

fractional distillation The separation of a mixture of liquids that have differing but similar boiling points. The fractionating column allows the separate liquids (or fractions) in a mixture to be collected at different temperatures. The temperature is higher in the lower regions of the fractionating column, which is where the less volatile compounds condense and are removed. The more volatile compounds progress up the column to condense at lower temperatures.

Frasch process The process in which sulfur is extracted from deep underground deposits. It consists of three concentric pipes that are sunk to the level of the deposit. Superheated water is forced down the outermost pipe and hot compressed air through the innermost pipe. As the steam melts the sulfur, it is forced up the middle pipe with air and water. Sulfur solidifies in large tanks on the surface.

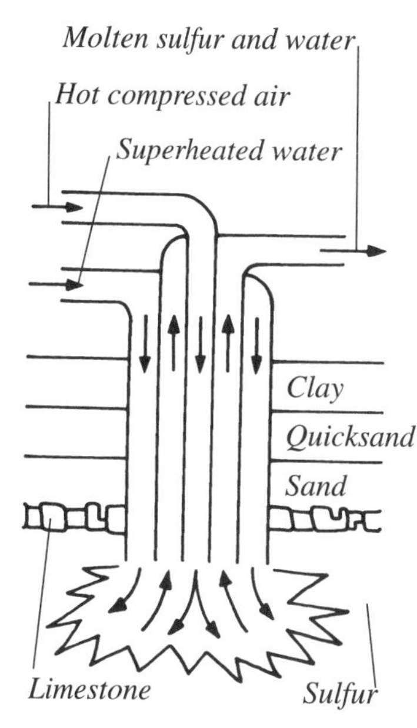

Frasch process

free fall Movement under the influence of gravitation that is not impeded by frictional resistance. Free-fall acceleration is constant.

free space An imaginary region containing no matter and no gravitational or electromagnetic fields, a temperature of absolute zero, and a refractive index of one. If light were to pass through free space, its speed would be maximal.

freezing The process by which a change of state from liquid to solid occurs. The freezing point is the temperature at which this change occurs (it is also the temperature of the melting point when the state changes from solid to liquid). It is the point at which the solid and liquid are in equilibrium. A freezing mixture is used to create a low temperature for chemical reactions. The mixture absorbs heat, producing lower temperatures than the original components of the mixture.

frequency The number of complete cycles of a periodic motion (such as the number of waves, or the number of swings of a pendulum, or the number of oscillations of electromagnetic radiation) passing a point in a given time. The unit is the hertz (1 cycle per second). Frequency is related to wavelength and the velocity of a wave, thus, frequency = velocity/wavelength.

frequency modulation (FM) A system of modulating a carrier wave used for transmissions in the VHF and microwave bands. A second wave motion is added to a carrier wave to carry the signal (sound or vision) to a receiver. The characteristics of this audio-frequency signal vary the frequency of the carrier wave without changing the amplitude. FM is inherently less susceptible to electrical interference than amplitude modulation (AM). *See also* amplitude modulation.

frequency spectrum The range of frequencies produced by different sources of sound.

Fresnel's biprism An isosceles prism with an angle of nearly 180°. It is used to produce interference fringes from the two refracted images of an illuminated slit.

Fresnel's mirrors Two plane mirrors inclined at an angle of almost 180°. They are used to produce interference fringes from the two reflected images of an illuminated slit.

friction The force that tends to oppose motion between surfaces in contact.

fulcrum The point at which a lever is supported. The lever turns around this point.

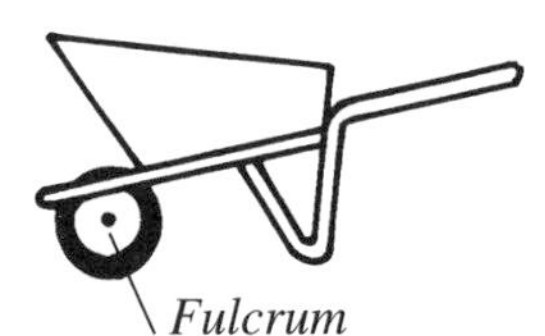

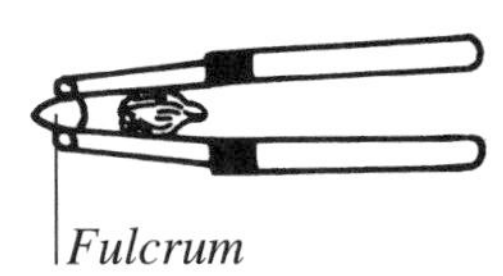

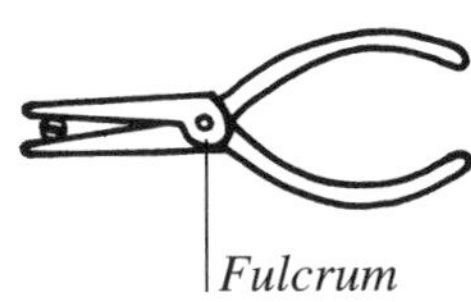

Fulcrum

function A relationship between variables such that a change in one will result in a corresponding change in the other. If a change in A results in a change in B, then B is the dependent variable and A is the independent variable.

fundamental (*harmonics*) The component having the lowest frequency in a complex vibration (fundamental frequency or first harmonic).

fundamental particles The large number of subatomic particles making up the matter in the universe. Electrons, protons, and neutrons were the first to be discovered, and many more have now been discovered. They are considered as variations of three types:
(1) leptons – electrons, muons, tauons – low mass particles;
(2) mesons – unstable, medium mass particles containing two quarks;
(3) baryons – proton, neutron – more massive particles containing three quarks. Quarks are the basic fundamental particles forming atomic nuclei.

fusion (1) (*melting*) The process by which a change of state from solid to liquid occurs.
(2) (*nuclear*) The process (which requires extremely high temperatures to initiate) by which two or more light atomic nuclei join, forming a single heavier nucleus. The products of fusion are lighter than the components. The mass lost is liberated as energy, given by Einstein's equation $E = mc^2$.

fusion energy *See* fusion.

fuzzy logic Computer logical processes that do not require precision or certainty in their data but can operate on degrees of truth or falsity. This approximates more closely to human logical processes and is of major interest to those dealing with artificial intelligence and neural networks.

gain The increase in voltage or power produced by an amplifier. The amplification of a signal, i.e., the ratio of signal output to signal input.

galaxy A large star system (millions of stars plus gas and dust) held together by gravitational attraction. Galaxies can be elliptical, spiral, or irregular in shape.

Galilean telescope The first astronomical refracting telescope. The eye lens is concave with a short focal length. The objective lens is convex with a long focal length. The separation between the two lenses is equal to the difference between the focal lengths of the lenses. It produces an erect final image.

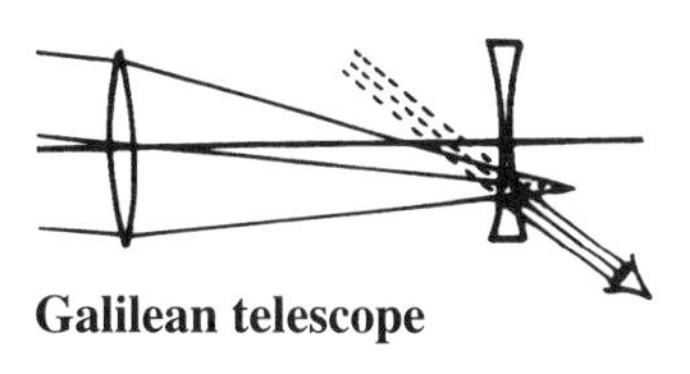
Galilean telescope

gallium arsenide A semiconductor compound used in the manufacture of high-speed, low-power consumption microelectronic integrated circuits. Gallium arsenide chips are more expensive to make than silicon chips. Ion-beam epitaxy is usually employed.

galvanizing The coating of iron or steel plates with a layer of zinc to protect against rusting. It is done either by dipping the iron or steel into a bath of molten zinc or by electrolysis.

galvanometer A sensitive instrument for measuring small electric currents. Most galvanometers are of the moving-coil type in which a coil of fine insulated wire can rotate between the poles of a strong permanent magnet. The current to be measured is passed through the coil and produces a magnetic field that interacts with the field of the permanent magnet to cause the coil to turn against the resistance of a light hair-spring. The coil carries a light pointer or a small mirror to deflect a beam of light across a scale ("weightless pointer"). The stronger the current, the greater the deflection of the pointer or beam.

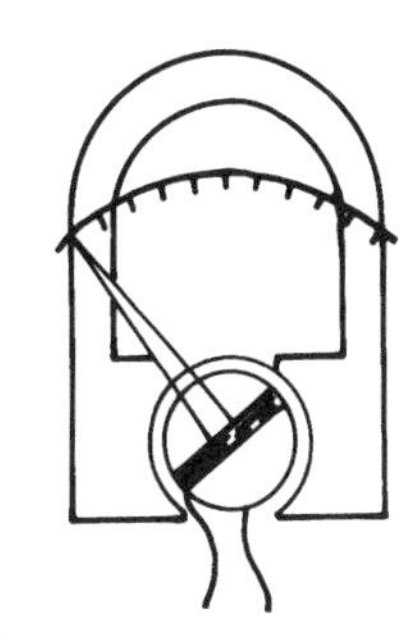
Galvanometer

gaussmeter An instrument for measuring the strength of a magnetic field.

gamma radiation (γ radiation) Very short-wave electromagnetic radiation emitted as a result of radioactive decay. It is the least ionizing and most penetrating of the three types of radiation emitted in radioactive decay. It will penetrate a thick metal sheet and is stopped only by over 15 cm of lead or by thick layers of concrete.

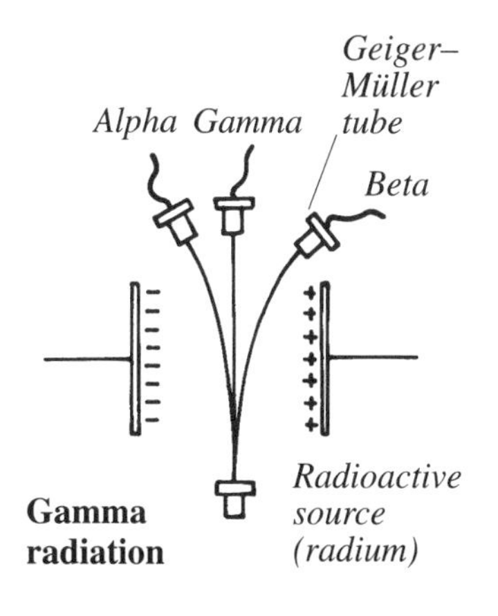

Gamma radiation

gas One of the three states of matter. In a gas, the particles can move freely through the space in which they are contained. Gas is the least dense of the states of matter.

gas law The equation combining Boyle's law and Charles' law, $PV = nRT$, where P is the pressure, V is the volume, n is the number of moles of gas present, T is the temperature measure in kelvin, and R is the universal gas constant.

gas phase The period when a substance is in gaseous form rather than a solid or liquid.

Gauss' law When a closed surface is in an electric field, the total electric flux at right angles to that surface is proportional to the sum of the electric charges within the surface.

Gauss' theorem A mathematical law defining a relationship between multiple integrals.

gear wheel A toothed wheel that transfers turning motion (torque) from one shaft to another. The number of teeth on the input and output wheels can be varied to change the gear ratio, depending on the requirements of either force or speed.

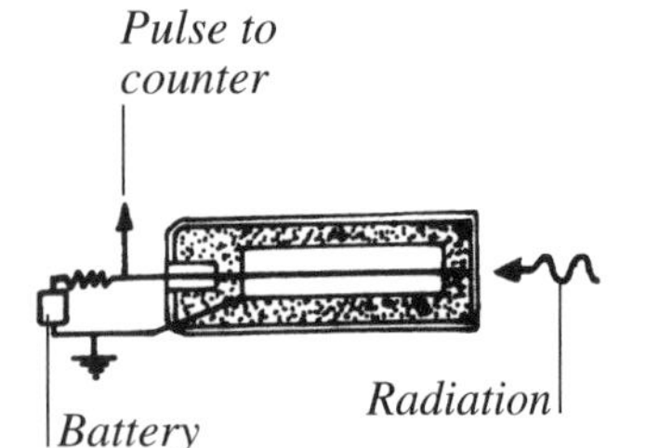

Geiger counter

Geiger counter An instrument used to detect and measure radiation capable of causing ionization (ionizing radiation). The Geiger counter consists of a sealed tube containing gas, such as neon or argon, and two electrodes with a potential difference of about 1,000 volts between them. If a quantum of radiation causes an ion to form in the gas, this ion, being a charged particle, is strongly attracted to one of the electrodes. Its velocity causes an avalanche of collisions and further ionizations, and a spike of current passes between the electrodes, causing an electrical pulse that can be heard as a sound in a loudspeaker or shown as a sudden brief deflection of the needle on an electric meter. Heavy ambient radiation causes a sustained sound or a persistent deflection of the meter, and this can be calibrated in units of received radiation. The counter was invented in 1908 by the German physicist Hans Geiger.

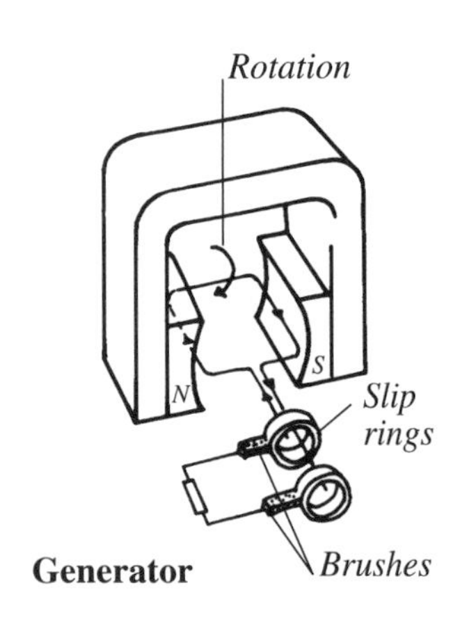

Generator

Geissler tube A tube that shows the lighting effect of the passage of an electric current between electrodes through ionized, low-pressure gas, and that can be used as a light source for spectroscopy. The tube was developed by the German mechanic Heinrich Geissler.

generator A machine that converts mechanical energy to electrical energy. At its most basic, it consists of an armature (coil of wire) rotated on a shaft in a field produced by a magnet. Electric current is induced in the coil by electromagnetic induction. The current is drawn from the coil by an arrangement of slip rings and carbon brushes.

generator stator coils The stationary coil around the armature in which the current is produced in a generator.

geometrical optics The geometry of light rays passing through optical systems and forming images.

geometric series A series of numbers with a constant ratio between successive terms, as in 1, 3, 9, 27, 81, 243. Compare arithmetic series, in which there is a constant difference between successive terms, as in 1, 4, 7, 10, 13.

geophysics The study of the physical properties of the Earth. Geophysics includes such subjects as geothermometry, seismology, glaciology, oceanography, geomagnetism, and geochronology.

geostationary orbit The orbit of a satellite that makes exactly one rotation of the Earth in the time taken for the Earth to rotate on its axis. If it is in the equatorial plane, the satellite will remain permanently above the same point on the Earth's surface and will appear to be stationary. Such a satellite will be at an altitude of 22,307 miles (35,900 km).

geothermal energy Power that is generated making use of the heat energy in rocks within the Earth's crust.

germanium The semiconductor element from which the earliest transistors were made. Germanium has been largely replaced by silicon.

Gibbs function or **Gibbs free energy** The energy absorbed or released in a reversible reaction at a constant temperature or pressure. It is calculated for a system from the enthalpy minus the product of the entropy and absolute temperature.

global positioning systems Satellite systems enabling location of points on the Earth's surface very accurately.

global warming Certain gases, such as CO_2, produced when fossil fuels are burnt and methane from animals and other sources form a layer in the Earth's atmosphere. This produces the same effect as glass in a greenhouse (*see* greenhouse effect). Short-wave infrared rays pass through the layer of carbon dioxide and methane and heat the Earth, which radiates longer-wave infrared radiation. This cannot pass the layer and is trapped in the Earth's atmosphere, leading to warming.

Gödel's theorem A proposition by American mathematician Kurt Gödel to the effect that in a formal axiomatic system, such as logic or mathematics, it is impossible to prove everything within the system without using methods from outside the system. All such logical systems must contain more statements than can be proved using their own set of rules. Some have taken Gödel's theorem to indicate that a computing system can never exceed human intelligence because its knowledge is restricted to the fixed set of axioms built in by the programmers, while humans can discover new concepts.

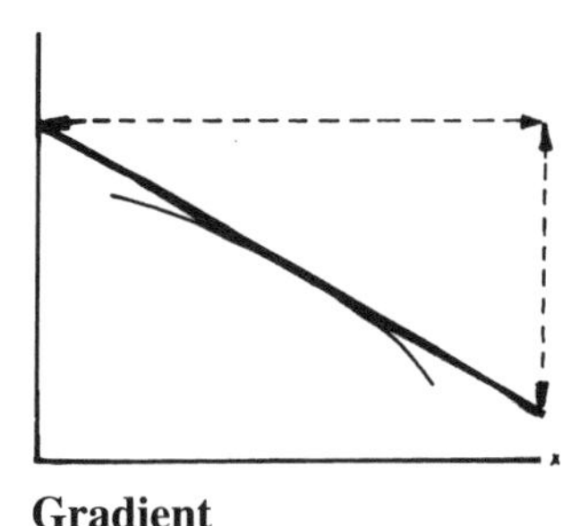

Gradient

gradient The slope of a tangent to a curve at a given point.

Graham's law The velocity with which a gas will diffuse is inversely proportional to the square root of its density.

grand unified theories Theories attempting to find a single combined explanation for electromagnetic, strong nuclear, and weak nuclear forces. A fully unified theory would also include gravitational force.

graphical user interface (GUI) A program interposed between the user and the operating system of a computer to provide a more "friendly" system of operation, usually involving screen symbols and a mouse. Commands are issued by moving the cursor to the desired symbol and "clicking" the mouse.

graticule or **reticule** A network of fine wires at the focal point of the eyepiece of a telescope or microscope. It is used as a reference point or for measurement.

grating spacing The distance between closely spaced lines used as a diffraction grating.

gravitational constant (G) The universal constant of gravitation used in Newton's law of gravitation.

gravitational field The area of space around a body in which it attracts another body because of their masses. The gravitational field strength gives a measure of the effect of a gravitational field on a mass and is measured by the force it exerts on any object placed in the field per unit mass.

gravitational mass The force exerted on a body by the Earth's magnetic field. It is measured by weighing the object. It gives the degree to which the mass being weighed is attracted to the Earth.

gravitational potential The change in potential energy for a unit mass that moves into an object's gravitational field.

gravitons Hypothetical particles that convey gravitational force in the way that photons convey electromagnetism.

gravity The effect of gravitational attraction between the Earth (or other heavenly body) and a body in its gravitational field. It is the strength of this field that is responsible for the weight of an object and the acceleration of free fall.

great attractor A hypothetical mass of great magnitude that is causing the Milky Way and associated galaxies to move toward it.

greenhouse effect Short-wavelength infrared radiation from the Sun passes through the glass of a greenhouse and warms the contents. These radiate longer-wavelength infrared radiation (the wavelength of the radiation depends on the temperature of the body emitting the radiation.) This radiation cannot pass through the glass, so the temperature in the greenhouse rises. The effect is used to describe how the Earth's atmosphere heats up.

Greenhouse effect

ground state The natural state of lowest energy of a particle, atom, or molecule.

group velocity The velocity of a group of electromagnetic waves passing through a medium that is not a vacuum. (This velocity is always less than the velocity of waves in a vacuum. Individual waves in a group appear to travel with a phase velocity that is greater than the velocity of waves in a vacuum.)

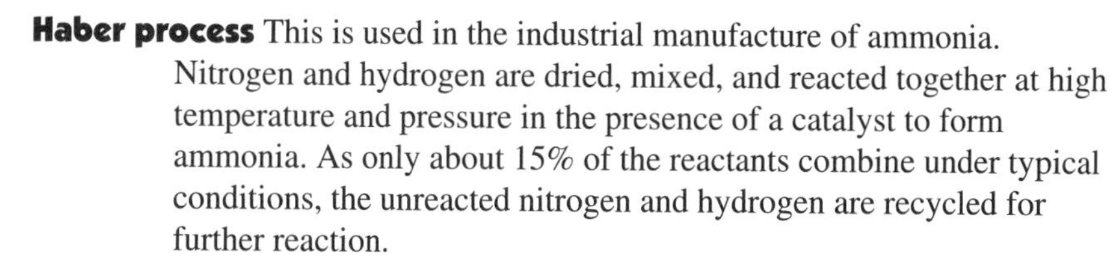

Haber process This is used in the industrial manufacture of ammonia. Nitrogen and hydrogen are dried, mixed, and reacted together at high temperature and pressure in the presence of a catalyst to form ammonia. As only about 15% of the reactants combine under typical conditions, the unreacted nitrogen and hydrogen are recycled for further reaction.

hadron An elementary particle that interacts strongly with other particles. Baryons (protons and neutrons) and mesons are hadrons.

half-life A substance that decays exponentially by the same ratio in equal intervals of time. The constant ratio is the half-life. The rate of radioactive decay of a substance is defined by its half-life.

Hall effect The production of a transverse voltage difference in a conductor, caused by the disturbance of the lines of electrical current flow, that occurs on the application of a magnetic field perpendicular to the direction of current flow. The Hall effect can be usefully applied in semiconductors.

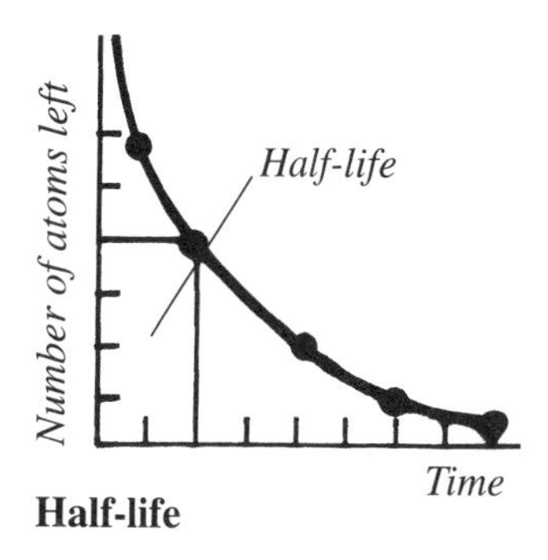

Half-life

Hall slice A thin slice of conducting material.

Hall voltage When a magnetic field is applied across a solid at right angles to the direction of the current, the moving electrons (from the current) move to one edge of the solid where they gather, causing a potential difference that can be measured.

halogens Elements of Group 7 of the periodic table. They are poisonous and nonmetallic. They all have seven electrons in their outermost shell, so they gain electrons easily to form univalent anions.

Hamilton's mill A device to illustrate the formation of electric wind at highly charged points.

hard disk A computer recording device consisting of one or, usually, more firm metal disks coated with magnetic material. The disks rotate at high speed on the same spindle, and each is provided with a radially moving read-write electromagnetic head that can thus cover the active area of its disk. The heads do not actually touch the magnetic surface but come very close to it, and their radial position can be set very quickly with high accuracy under the control of the disk controller circuit and the program. Currently, typical capacities of hard disk drives in personal computers are between 1,000 megabytes (1 gigabyte) and about 20 gigabytes.

hardness *See* Mohs' scale.

hardness in water The presence in water of calcium and magnesium ions, which restrict the ability of soap to form a lather and leave deposits in pipes. There are two types. In temporary hardness, soluble hydrogen-carbonate compounds of calcium, magnesium, and iron are present. When the water is heated, these salts form insoluble carbonates that are precipitated. Permanent hardness is caused largely by calcium sulfate and cannot be removed by boiling the water.

hardware The physical machinery of a computer, incapable of doing anything by itself. *Compare* software, the programs that instruct the hardware and prompt it into action. When a computer is turned on, it immediately reads some of its software and carries out certain initial instructions. It then waits for further instructions from the keyboard or other input device.

harmonic The part of a complex vibration that is a simple multiple of the fundamental. Harmonics are often termed *overtones*; they are counted in order of their frequency above (but excluding) the fundamental, e.g., the third harmonic is the second overtone.

heapsort A computer sorting algorithm that uses the information gained in one stage to save on subsequent comparisons (tree selection).

heat When energy is transferred between a system and its surroundings because of a temperature difference, the energy is heat.

heat capacity The heat capacity of an object is the product of its mass and its specific heat capacity.

heat energy A system has heat energy because of the kinetic energy of its atoms and molecules (due to translation, rotation, and vibration). It is transferred by conduction, convection, and radiation.

heat flow rate The rate at which heat energy is transferred through a medium by the vibration of its atoms.

heat of reaction The difference between the enthalpy of the products of a reaction and the enthalpy of the reactants. (The heat of reaction is negative if the reaction is exothermic and positive if the reaction is endothermic.)

heat pump A closed system containing a working fluid that takes heat from an external source and is then mechanically compressed, resulting in a further rise in the fluid's temperature. The fluid then passes through a heat exchanger and evaporates, cooling further, and is then passed out to gain heat from the external source.

helical motion Moving in a spiral.

heterodyne principle A design feature of almost all radio and television receivers in which the incoming signal is mixed with a signal from a local oscillator that is tuned along with the incoming signal so that there is a constant difference between the two frequencies. The resulting difference frequency is amplified by the intermediate frequency amplifier before being demodulated. The advantage is that the bulk of the amplification takes place at a fixed low frequency – the intermediate frequency – rather than at a wide range of frequencies, which would require all the stages to be tuned simultaneously.

heuristics A nonalgorithmic set of rules for solving problems in which educated guesses and a "trial-and-error" approach replace the application of established knowledge. Heuristics is of great interest to workers in the field of artificial intelligence.

high-pass filter An electronic device that discriminates between different frequencies of current, allowing those above a given frequency to pass with little hindrance (low impedance), while offering a high impedance to those below the particular frequency.

high-temperature materials Materials, such as certain metals, alloys, and ceramics, capable of retaining their physical properties at temperatures above about 540°C (1,000°F). Such materials are needed for use in furnaces, gas turbines, aircraft jet engines, and nuclear reactors. They must be inherently strong and resistant to oxidation.

highway *See* bus.

holes Vacant electron energy states near the top of energy bands in solids. The absence of the negative electron is equivalent to a positive charge, and holes behave exactly as if they were positive charges and can move through solids in the opposite direction to the movement of electrons. Hole conduction is an essential feature of semiconductor action and can be promoted by "doping" semiconductors with a very small proportion of donor atoms with a lower vacancy than the main material.

holographic memory A computer memory store in the form of a hologram that is read by a low-power laser. Such memories can securely store large amounts of data in a small physical space.

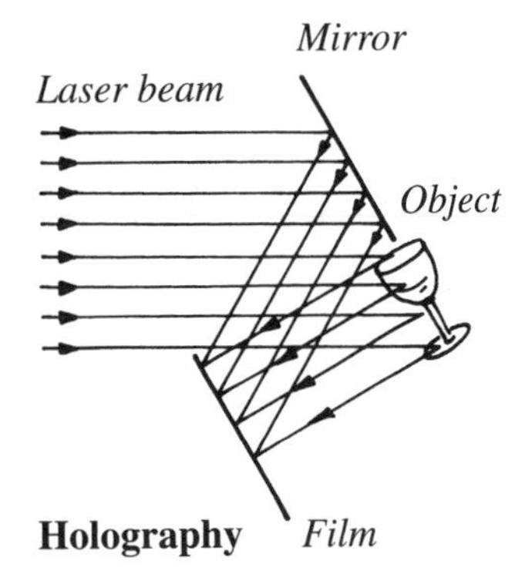

Holography

holography A technique for making three-dimensional pictures using laser light. The beam from the laser is split into two parts, one part illuminating an object. A photographic film receives light from the object, and the second part of the beam, from the laser. This produces an interference pattern on the film. When laser light is shone on this film, it produces two images that give a three-dimensional effect.

homologous series A series of related organic compounds. The formula of each member differs from the preceding member by the addition of a $–CH_2–$ group. Each series has a general formula; for example, the general formula for alkanes is C_nH_{2n+2}. The properties of each series, though similar, change gradually and regularly with increasing molecular weight.

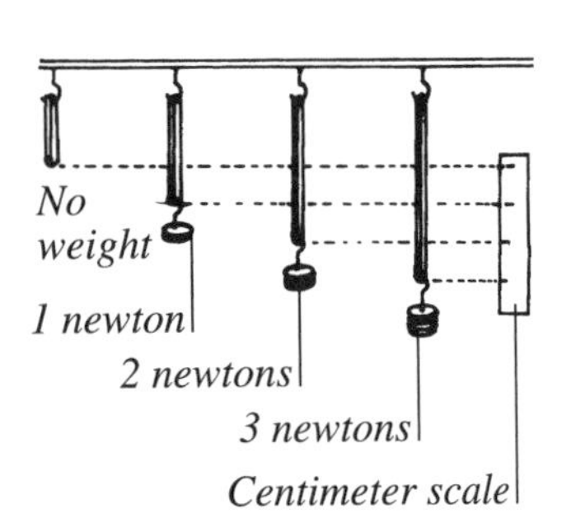

Hooke's law

Hooke's law The extension of an elastic material is directly proportional to the force that causes the extension, provided that the elastic limit is not exceeded.

Hubble's constant A figure based on the assumption that there is a direct relationship between the distance of a celestial object and its recessional velocity. The velocity divided by the distance should always give the same number, known as Hubble's constant after American astronomer Edwin Powell Hubble whose work on the recession of the galaxies led to the concept of the expanding universe. Hubble's constant is not universally agreed, but widely accepted values are 15 and 29 km per second per million light-years.

If Hubble's figure is a constant, then the age of the universe is its reciprocal. Other estimates, from geological evidence, especially of the amounts of radioactive elements, generally agree with these figures. The most commonly held age for the universe is about 15 billion years.

Huygens' principle Each point on a wavefront acts as a source of secondary wavelets.

hydrated crystals Crystals of a salt that contain water of crystalization.

hydraulic jack A machine for raising heavy weights over short distances. The lifting part is connected to a piston in a cylinder filled with fluid, the pressure of which is controlled by a pump.

hydraulics The science that relates to the flow of fluids.

hydride A chemical compound formed between hydrogen and another element in which the hydrogen is present as the hydride ion (H^-). This happens with strongly electropositive elements. The electrolysis of hydrides liberates hydrogen at the anode.

hydrocarbons Organic compounds that contain carbon and hydrogen only. They can be aliphatic or aromatic, saturated or unsaturated.

hydrodynamics A branch of dynamics: the study of the motion produced in fluids by applied forces.

hydrogenation A chemical reaction adding hydrogen to another molecule (usually organic) in the presence of a catalyst. Hydrogenation reactions often involve the adding of hydrogen to unsaturated molecules.

hydrogen bomb An atomic weapon in which a fusion reaction occurs in heavy hydrogen (deuterium) as a result of a triggering fission reaction in uranium or plutonium. For fusion to be self-sustaining, a temperature of about 350 million K is necessary, and the only practicable way to achieve this in a weapon is to explode a conventional fission atomic bomb. At this temperature the energy released by the initial fission is sufficient to maintain the temperature in the heavy hydrogen and initiate a fusion chain reaction.

hydrolysis The interaction of water with a salt to form an acid and a base. The water dissociates to H^+ and OH^- ions.

hydrosphere That part of the Earth's surface composed of water.

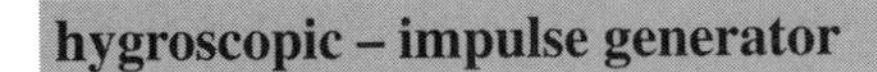

Start with white copper sulfate

After several minutes, blue copper sulfate forming

After several hours, no further change

Hygroscopic

hygroscopic A substance that absorbs moisture from the air without becoming liquid.

hysteresis The phenomenon in which the effect lags behind the cause of the effect. Hysteresis is common in science, but the best-known example is when the magnetism induced into a ferromagnetic material lags behind the changes in the external field causing the magnetism. In general, if the effect Y is plotted against the cause X, a closed curve, known as the hysteresis loop, is formed.

ideal gas A gas that obeys the gas laws and one in which molecules have negligible volume and the forces of molecular attraction are negligible. No known gas is completely ideal.

ignite To heat a mixture of gases to the temperature at which combustion occurs.

illuminance The quantity of light falling on a unit area of a surface (it is inversely proportional to the square of the distance of the surface from the light source).

imaging The process of obtaining images of Earth and objects in space. Images are obtained using different parts of the electromagnetic spectrum and are then either recorded or transmitted directly to Earth.

immiscible Substances that do not mix and form more than one phase when brought together.

impedance (Z) The total resistance to alternating current flow in a circuit. It is the sum of the resistance and the reactance.

impedance match The condition that ensures the maximum transmission of power, when a source of alternating current is connected to a load. Impedance matching is especially important when the impedance of the source is high and a low load impedance would result in heavy losses. Maximal power transfer occurs at any frequency when the two impedances are equal. Maximum voltage transfer occurs when the impedance of the load is very high compared to that of the source.

impulse A force that acts over a very short time. For example, when two bodies collide, the impulse of the force is equal to the change of momentum produced in either body.

Impulse

impulse generator A device that delivers a short surge of electrical power, usually for testing insulators or the effect of such surges on electrical equipment, or for other purposes, such as the magnetization of permanent magnets. It consists of a high-value capacitor that can be charged slowly and discharged quickly through a low-resistance circuit.

impurity (1) A substance that is present in another material but is not wanted and must be removed.
(2) Small amounts of an element added to a semiconductor to alter the properties of conduction and conductivity of solid-state devices.

incandescence The emission of visible light by a hot body. For a dark-adapted eye this occurs at a temperature of about 390°C (730°F) as a colorless glow. Dull red incandescence occurs at a temperature of about 500°C (930°F).

incandescent lamp An electric light bulb that produces light by heating a filament, usually of tungsten, to a temperature at which yellow or white light is produced. Infrared and ultraviolet light are also produced, thus reducing the efficiency of the device as a visible light source. The filament is enclosed in an evacuated bulb or a bulb filled with inert gas so as to avoid oxidation and is usually in the form of a coil or a coiled coil. The electrical resistance of the filament varies with the supply voltage and the light output.

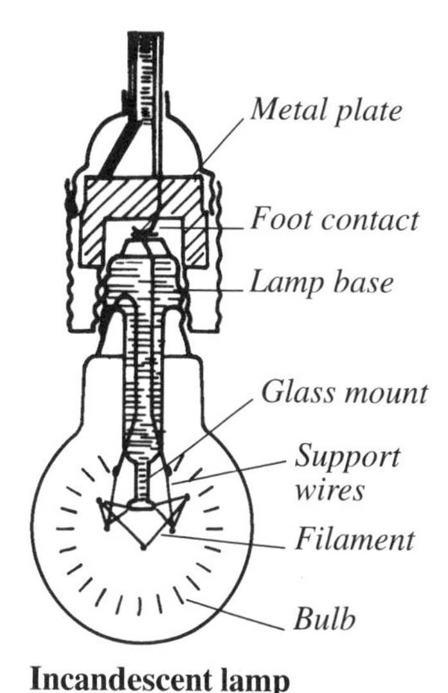

Incandescent lamp

incident ray A ray that strikes a reflecting or refracting surface.

indicator A substance that indicates, by a change in its color, the degree of acidity or alkalinity of a solution or the presence of a given substance.

induced current Current produced by the induced e.m.f. during electromagnetic induction.

inductance The extent to which an element of a circuit forms a magnetic field when carrying an electric current.

induction The change in a body's electric or magnetic state caused by its nearness to a field. Electrostatic induction is an electric charge produced on a conductor if it is placed in an electric field surrounding a charged body; it causes movement in the surface electrons on the conductor.

induction coil A form of electrical transformer used to produce a high-voltage, low-current alternating current from a low-voltage source of direct current. The source current is regularly interrupted by a switch of some kind, usually mechanical, and flows in the primary winding, which consists of a relatively small number of turns of thick insulated wire wrapped around a soft iron core. The secondary winding consists of a large number of turns of thinner insulated wire wrapped around the primary winding. The spark plugs of car engines are supplied by an induction coil through the distributor.

induction heating The use of induced eddy currents to heat a material that can conduct electricity. A coil, through which a large alternating current flows, is placed closely around the material to be heated. Induction heating is used extensively in the metalworking industry to melt metals and produce alloys.

induction motor An alternating current electric motor in which the current in the secondary winding (rotor or stator) is caused to flow by induction from the magnetic field caused by the current flowing in the primary winding.

inertia Reluctance to move. The tendency of a body to oppose a change in its motion. The inertia of a body is proportional to its mass. The inertial mass of a body is determined by its acceleration when a standard force is applied.

infinity (1) (*optics*) A point that is so far away from an optical system (lens, mirror) that the light emitted by it falls in parallel rays on the system. (2) (*physics*) A quantity of such a size that any variation in its dimensions leaves it unchanged.

information processing The derivation of information from other information or the reorganization or reallocation, to different sites, of information. Information processing is the chief function of the computer and the chief occupation of people who use computers.

infrared Electromagnetic radiation with a greater wavelength than the red end of the visible spectrum (.75–1,000 μm). Infrared radiation lies between the visible and microwave regions of the spectrum of electromagnetic radiation.

inorganic Compounds that do not contain carbon, with the exceptions of carbon monoxide, carbon dioxide, carbonic acid, and carbonates.

input device Anything by which data can be transferred to a computer, such as a keyboard, bar code reader, data tablet, document reader, magnetic card reader, speech recognition device, or scanner.

insoluble A substance that does not dissolve in a particular solvent under certain conditions of temperature and pressure.

instruction set The list of computer-program machine-code instructions that a microprocessor can recognize and act on.

insulation The process of inhibiting the flow of energy (electrical, heat, sound, vibration) by placing nonconductors or insulators in its path.

integrated circuit (IC) A complete electronic circuit on a silicon or other chip, often containing many thousands, or even millions, of active units (transistors) and the associated resistors and capacitors. Integrated circuits may range from a very simple circuit, such as a voltage amplifier or a digital gate, to the nearly complete electronics of a computer.

interference The interaction between waves of the same frequency emitted from different sources. The principle of superposition governs the wave in which the wavefronts combine, making interference patterns.

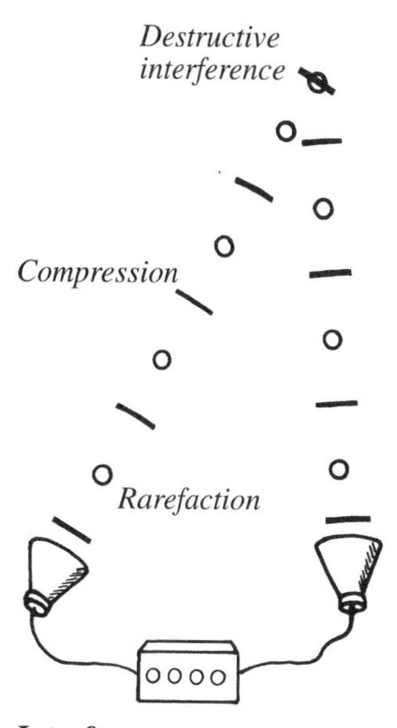

Interference

interference fringes The variation in disturbances produced by waves of the same frequency from different sources; they appear as alternate light and dark bands.

internal combustion engine An engine in which energy is produced in a cylinder by the combustion of a gas-and-air mixture. This energy is converted to mechanical energy by driving pistons in the cylinders.

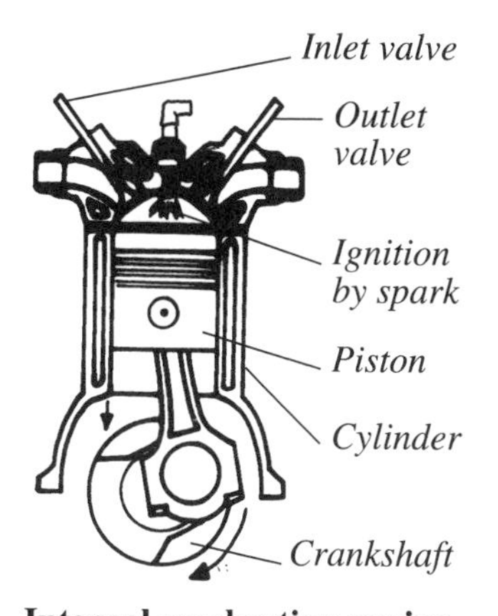

Internal combustion engine

internal energy The total energy a body has as a result of the relative motions and kinetic and potential energies of all its component particles. It is indicated by the temperature of a body.

interrupt I/O A method of controlling the central processing unit (CPU) of a computer when data must be moved in or out. A signal is sent to the CPU and this sets a program interrupt, which, at an appropriate time, determined by various priorities, causes the CPU to interrupt the program so that data can be moved in or out.

inverse-square law The law that states that the intensity of radiation, such as light, decreases with the square of the distance from the source of radiation.

inverse-square relationship A property of a system that varies inversely with the square of another property of the system.

inverting amplifier An amplifier whose output is the inverse of its input.

ion An electrically charged atom or group of atoms. An atom or group of atoms that has lost one or more electrons, so it is positively charged, or gained one or more electrons, so it is negatively charged.

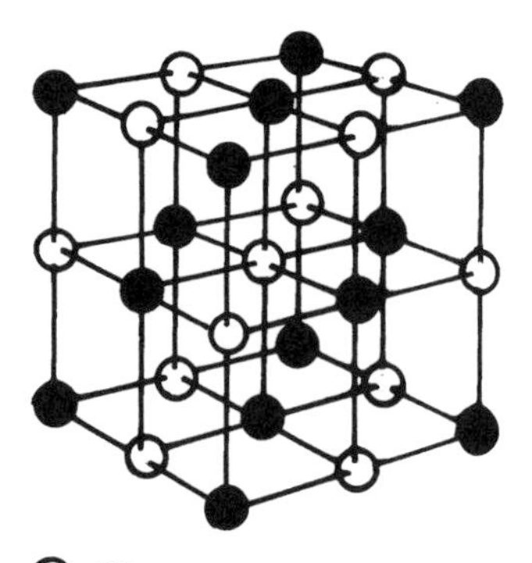

○ Cl^-
● Na^+
Ionic lattice

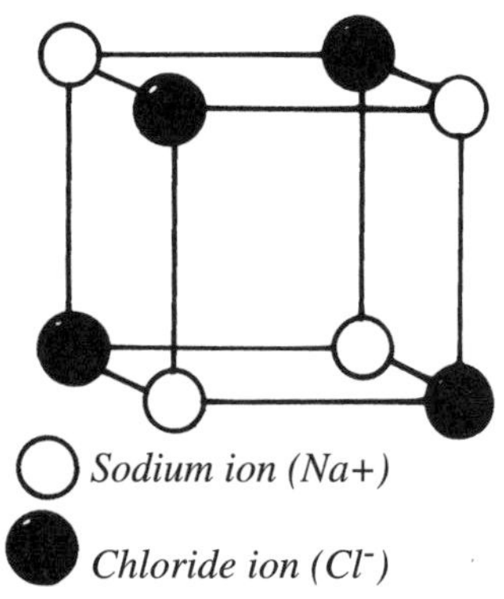

Ionic lattice

ionic bonding *See* bonding.

ionic lattice An ionic crystal of two or more elements that is held together by the electric forces (ionic bonds) between negative and positive ions in a regular structure.

ionic radius The approximate distance between the center of an ion's nucleus and the edge of its surrounding electron cloud in a crystal.

ion implantation A method of adding materials to the surface region of other materials by bombarding them with a high-energy beam of ions. This can be done with great precision as to location, area, and depth, and is an important procedure in the manufacture of semiconductor devices.

ionization energy The energy needed to remove completely an electron from a neutral gaseous atom or ion against the attraction of the nucleus. The energy needed to remove the first electron is the first ionization energy, the energy for the second is the second ionization energy. These become progressively larger.

ionizing particles Charged particles (α, β) that cause ionization in a medium when passing through it.

ionizing radiation Any electromagnetic or particle radiation that causes ionization in a medium when passing through it.

ionosphere Part of the Earth's upper atmosphere where ions and free electrons are found.

ion propulsion A form of vehicular propulsion for artificial satellites and spaceships in which ions of various substances are accelerated to a high speed by an electric field and driven out of the rear of the vehicle so as to produce forward motion by reaction. Power for the purpose can be obtained from the Sun, using solar-electric generators, or from a nuclear reactor. The method is very economical on materials, which is an important consideration in space.

isotope Atoms of the same element (all chemically identical) having the same atomic number but containing different numbers of neutrons, giving different mass numbers. Some elements occur naturally as a mixture of different isotopes. All elements have artificially produced radioisotopes.

Josephson junction A superconducting device, consisting of a junction between two metals, that exhibits controllable electron tunneling. The Josephson junction can be used as a very fast electronic switch, changing from "on" to "off" in less than 1 nanosecond and requiring minimal power. Computer hardware based on this principle must be immersed in liquid nitrogen or liquid helium, but the speed of operation will often justify this.

joule The unit of work, energy, and heat. In electrical terms, the joule is the watt-second. One hundred joules are needed to run a 100W bulb for one second. Although defined in terms of work, the joule is a unit of heat energy and is used increasingly to replace the calorie in nutritional contexts. The calorie is the amount of heat required to raise 1 g of water by 1°C, but in nutrition, the calorie is taken to be one thousand times this figure and is called a kilocalorie. A (kilo)calorie is equal to 4.2 (kilo)joules.

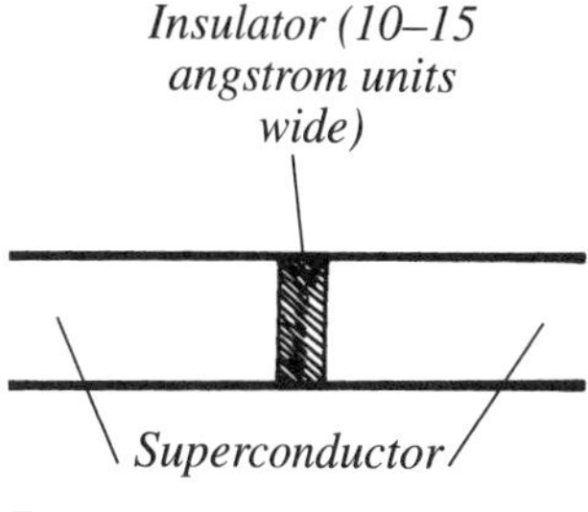

Josephson junction

Joule's law The relationship between the electric current flowing in a conductor and the heat produced as a result. Joule's law states that the rate of heat produced (in watts) is proportional to the resistance of the conductor (in ohms) multiplied by the square of the current (in amps). The law was first stated by the English physicist James Prescott Joule.

Joule-Thomson effect The decrease in temperature that occurs if a gas is expanded adiabatically through a small aperture. The reason for this is that energy is used to overcome cohesion between the gas molecules. The effect is the basis of most domestic refrigerators.

junction detector A detector of ionizing radiation using a reverse-biassed semiconductor junction. Reverse bias normally prevents current flow, but the ionizing effect of radiation allows a current that is directly proportional to the strength of the radiation to pass. The junction detector is one of the most important quantitative methods of radiation detection and measurement and has wide applications in space science, astronomy, electron microscopy, and medicine.

kaon or **kappa meson** A subatomic particle of medium weight, a hadron, that experiences a strong nuclear force.

Karnaugh map A pictorial representation of Boolean truth tables done as an aid to simplification or minimalization.

kb *See* kilobyte.

Kelvin scale A temperature scale that has no negative values. Its lower fixed point is absolute zero. The size of the unit, the kelvin, is the same as the degree Celsius, and the triple point of water is 273.16K.

Kepler's laws of planetary motion (1) Each planetary orbit is an ellipse with the Sun at one focus.
(2) A line from the Sun (radius vector) to the planet covers equal areas in equal times.
(3) The square of a planet's orbital period (year) is proportional to the cube of its mean distance from the Sun.

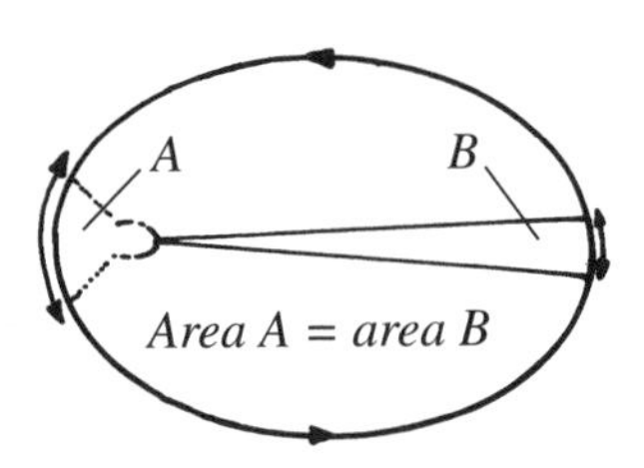

Kepler's laws

kilobyte (kb) A widely used unit of size or capacity for computer files or memories. In a decimal system, *kilo* means 1,000. But in a binary system, a kilo is 2 to the 10th power, which is 1,024. A kilobyte is thus 1,024 bytes, and a megabyte is 1,024,000 bytes.

kinetic energy The energy that a body possesses because of its motion. If a mass (m) is moving with velocity (v), its energy (E) is given by $E = \frac{1}{2}mv^2$.

kinetic theory of gases A model that explains the bulk properties of a gas by considering the movement of the molecules in the gas. It is considered that:
(1) the gas molecules are elastic spheres (pressure is hence caused by their bombardment of the walls of the container);
(2) the size of the molecules is small compared with their spacing;
(3) molecules do not exert forces on each other unless they are colliding.

kinetic theory of matter The theory that the particles of all matter, gas, liquid, and solid, are in a state of constant, vigorous, temperature-related motion. The theory is most readily appreciated in the case of gases, where it explains the observable facts about gases and the gas laws. In solids, particle movement is known as lattice vibration.

Kirchhoff's laws (1) At a junction in a circuit, the algebraic sum of the currents is zero.
(2) In a closed circuit, the algebraic sum of the electromotive forces is equal to the algebraic sum of the products of currents and resistances.
The laws were first enunciated by German physicist Gustav Robert Kirchhoff.

K-mesons Subatomic particles believed to be responsible for holding together the protons in the atomic nucleus in spite of the mutual repulsion of their like charges. Like charges repel; unlike charges attract. A powerful force is needed to prevent such closely aggregated like charges from flying apart.

large-scale integration (LSI) Chip technology in which at least 10,000 transistors and associated components are fabricated onto a single silicon chip. Modern LSI commonly involves more than a million transistors or other active devices on a single chip.

laser Acronym for *l*ight *a*mplification by *s*timulated *e*mission of *r*adiation, a device that can produce a narrow beam of intense monochromatic light that has high energy and little spread (the rays are parallel). Laser light is coherent – all rays are in phase with each other. This allows focusing to a much finer point than is possible with noncoherent light.

laser Doppler velocimeter An instrument that uses a laser and the Doppler effect to measure the velocity of particles in a fluid.

laser photochemistry The use of lasers to induce or affect chemical reactions.

laser spectroscopy Spectroscopy using laser light rather than noncoherent light. This has improved the resolution and extended the value of a wide range of spectroscopic techniques.

latch An electronic circuit that can store a single binary digit (bit) until the next clock pulse arrives.

latent heat The quantity of heat that is absorbed or released by a substance during a change of state (fusion or vaporization) at constant temperature.

latent image The undiscernible image on a photographic emulsion before development.

lattice vibration Periodic, heat-related oscillations of the atoms in a crystal lattice about their mean positions. The amplitude of the vibration increases with temperature until, at the melting point of the solid, the atoms leave their lattice sites. Atomic movement decreases with lowering temperature but never completely ceases, even at absolute zero. The residual vibration at absolute zero is called zero-point vibration and is a consequence of the uncertainty principle of quantum mechanics. Lattice vibration is the means by which heat is conducted through a solid. The scattering of conduction electrons in metals by lattice vibration results in the increase in the electrical resistance that is observed with a rise in temperature and the decrease in resistance that is observed with a fall in temperature.

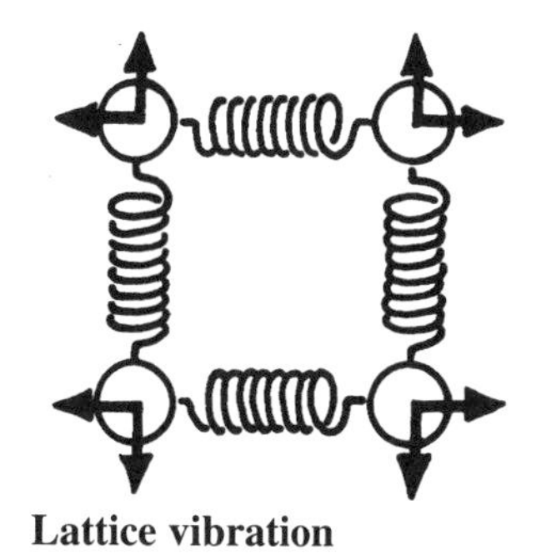

Lattice vibration

law A rule describing certain natural observable phenomena or the relationship between effects of variable quantities. Scientific laws are derived from large numbers of observations of a particular phenomenon and the failure to observe any instances in which, under the described conditions, the phenomenon does not occur. The logic behind scientific laws is inductive, and they are all provisional.

law of conservation of energy Energy may be converted from one form to another but it can be neither created nor destroyed in a closed system.

law of conservation of matter (mass) Matter can be neither created nor destroyed in a closed system.

law of conservation of momentum The total momentum of colliding bodies is the same before and after impact, assuming there is no external force applied.

law of moments For a body in equilibrium being acted on by forces, the sum of the clockwise moments about any point is equal to the sum of the counterclockwise moments about the same point.

laws of reflection When an incident light ray is reflected at a surface it is reflected in the same plane as the incident ray and the normal (perpendicular to the surface at the point of reflection). The angle of reflection is equal to the angle of incidence.

laws of refraction (1) When an incident ray is refracted at a surface, it is refracted in the same plane as the incident ray and the normal. (2) The ratio of the sine of the angle of incidence to the sine of the angle of refraction is a constant for a given pair of media – Snell's law.

LCD *See* liquid crystal.

LDR Abbreviation for light-dependent resistor, an electronic device whose resistance decreases when light falls on it.

leaching Losing soluble substances in a liquid that flows through the substance.

Leclanché cell A primary cell with an e.m.f. of 1.5 volts and internal resistance of 1 ohm. The positive electrode is a carbon rod, which is surrounded by a mixture of powdered carbon and manganese dioxide in a porous pot. The pot stands in ammonium chloride solution contained in a zinc pot, which forms the negative electrode.

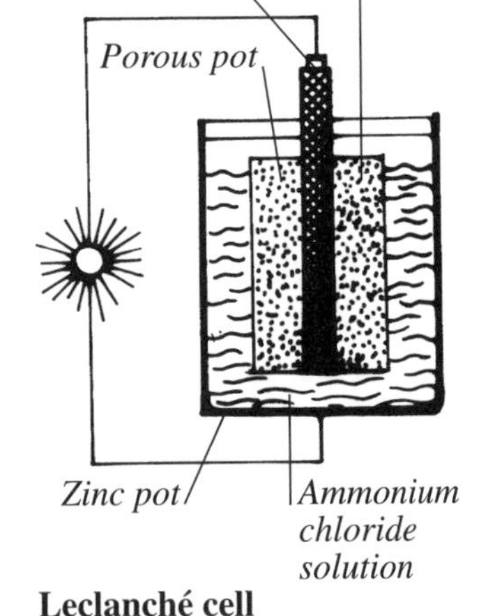

Leclanché cell

LED Abbreviation for light-emitting diode, a diode that emits light when a current passes through it. The color of the light emitted depends on the material used.

lens A shaped piece of transparent material that bends light by refraction. The shape of the surface determines the way in which the light is refracted. A convex lens causes parallel incident light to focus at a point; a concave lens causes light rays to diverge.

Lenz's law The principle that when an electromotive force is induced in a conductor it is always in such a direction that the current it produces will oppose the change that caused the electromagnetic force. This change is commonly the movement of a conductor through a magnetic field, so the current will produce a force that opposes the movement. Alternatively, the change may be the movement of a magnetic field in relation to a conductor. In this case, the current will produce an opposing magnetic field. The law was first stated by German physicist H. F. E. Lenz.

lepton A low-mass subatomic particle. Electrons, muons, neutrinos, and photons are leptons.

lever A simple machine consisting of a rigid beam turning about a pivot. A force is applied to the beam to overcome a load at another point on the beam. There are three types of levers, depending on the relative positions of pivot, force, and load on the beam.

Lever

lever balance An instrument for measuring mass using the law of moments. A mass is placed at one end of a lever, causing a turning moment about a pivot. The degree of turning is indicated on a calibrated scale at the other end of the lever.

lift *See* Bernouilli effect.

ligand An atom (or group of atoms) surrounding the central atom in a complex.

light amplifier An image intensifier that produces an image of more intense brightness than that seen by the unaided and dark-adapted eye.

light-emitting diode A small electronic device consisting of a semiconductor junction that produces a cold light, usually red, when a small electric current passes through it, i.e., when it is forward biassed. LEDs are widely used as "on" indicators and, because they can produce a very small point of light, can be combined to form displays of alphanumeric characters or more complex image displays. They are cheap, long-lived, and have very small current and voltage requirements, although greater than those of liquid crystal displays.

light energy Electromagnetic radiation that is detected by the eye. The wavelengths are between 400 (violet) and 730 (red) nanometers.

lightning An electrical discharge in the atmosphere. Energy is transferred to light and thermal energy. This causes a sudden expansion of air, which causes thunder.

light-year Unit of length. It is the distance traveled by light in a vacuum (space) in a year: 9.46×10^{12} km, or 63,240 astronomical units.

limewater A solution of calcium hydroxide that is used to test for the presence of carbon dioxide. If carbon dioxide is bubbled through limewater, a solid precipitate of calcium carbonate is formed.

linear expansivity The increase in length per unit length of an object when its temperature rises by one degree.

linear relationship The relationship of two quantities when a change in one causes a directly proportional arithmetical change in the other. The term *linearity* is often applied to electronic amplifiers to describe the uniformity of the degree of amplification over the working frequency range. The term describes the fact that if the two quantities are plotted on a graph, the result will be a straight line.

liquid A state of matter between solid and gas. Particles are loosely bonded, so can move relatively freely. A liquid has low compressibility.

liquid crystal A liquid that has some crystal properties. Liquid crystals are nematic substances – substances that exist in an intermediate state between liquid and solid (mesomorphic) in which a linear orientation of the molecules changes their physical properties. The application of an electric field causes crystals to realign so as to affect light shining on the crystal in different ways; one looks light, the other dark. Liquid crystals are used in digital displays (liquid crystal displays, LCDs), which may be monochrome or color, and are often backlit so as to make them more legible in bright conditions. Color displays may be of the double supertwist nematic type or the thin film transistor type. The former is cheap but reacts slowly and does not give particularly sharp colors. The latter produces an excellent display but is expensive.

Lissajous' figures Patterns formed on an oscilloscope screen when the electron beam is moved simultaneously by the electric fields of two sinusoidal voltages at right angles to each other. If the frequencies are equal, a single closed figure, such as a circle, is formed on the screen. If one is twice that of the other, a figure eight is formed. Three loops are formed if one frequency is three times that of the other, and so on. As long as one frequency is an integral multiple of the other, a close figure with multiple loops is formed. Lissajous' figures are a useful way of comparing the frequencies of two sinusoidal voltages applied to the plates of an oscilloscope or to a Lissajous' computer program. The idea was proposed by the French physicist Jules A. Lissajous who used vibrating mirrors.

local area network (LAN) A way of connecting personal computers within a building or small area so that they can communicate with each other and with a central file server computer with large storage capacity, from which they can derive data. Networking involves special problems and requires sophisticated software.

Lodge's tube A glass tube used in electrolysis.

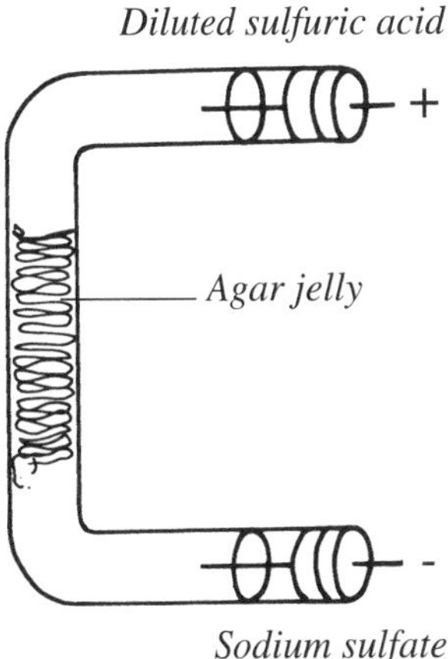

Lodge's tube

logic gate An electronic circuit that processes information coded in terms of voltage levels, usually representing 1s and 0s, or "true" or "false." The logic is implemented by three basic gates, the AND, OR, and NOT gates, or the derivatives of these, the NAND (not AND) and NOR (not OR) gates. Complexes of such gates make adders, multipliers, and so on.

logic state The state of truth or falsity of a binary signal. Only these two states are allowed in Boolean logic, and nearly all computers still operate on this basis, using logic circuits. "Fuzzy" logic, allowing a range of values between true and false, corresponds more closely to how biological systems operate and is used in neural networks and other attempts to implement artificial intelligence.

log in The act by which a computer user identifies himself or herself to the machine and is thus authorized to have access. Log in usually involves a form of authentication, often by typing in a recognized password. Log in implies that at the end of the session, the user will log out.

lone pair (electrons) A pair of electrons in the outermost shell of an atom that are not used to make covalent bonds with other atoms.

longitudinal waves A wave motion whose displacement is in the direction of its propagation, for example, a sound wave that travels through air as a series of compressions and rarefactions.

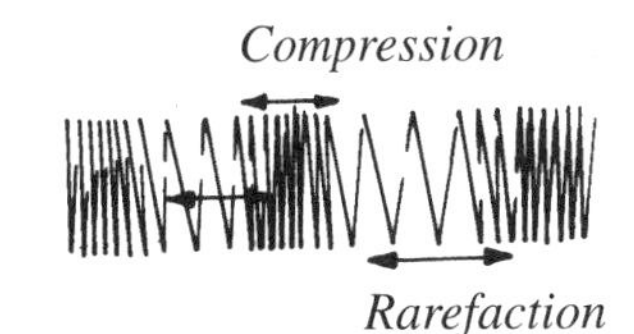

Longitudinal wave

loudness The subjective experience of the intensity of sound or the amplitude of sound waves. Because the human ear varies in sensitivity at different frequencies, loudness is not the same for different frequencies at the same sound amplitude. Sensitivity is highest between 1,000 and 4,000 Hz and is lower above and below this range. Loudness is compared by using a logarithmic scale that is adjusted to take into account these differences in sensitivity.

loudness level The comparison of the frequency of a specified sound with a reference tone. The comparison is made using decibel units. These are, however, units of power comparison, not units of loudness.

luminescence Light emission other than that caused by purely thermal energy. Light may be produced by part of the energy released from slow chemical reactions (chemoluminescence); it may be produced by the exciting effect of various forms of radiation (radioluminescence or photoluminescence; or it may be produced by the presence of an electric field (electroluminescence).

luminous energy Energy radiated in the visible part of the electromagnetic spectrum, specifically at a wavelength between 380 and 760 nanometers.

luminous flux The amount of light that passes through an area in one second.

lycopodium powder A fine powder obtained from the spores of plants belonging to the Lycopodium group.

Lyman series A series of lines in the ultraviolet part of the hydrogen emission spectrum.

machine code The software that provides immediate instructions to a computer's central processing unit in a form that can immediately be "understood" by the CPU. Machine code is the lowest-level programming language and consists simply of a string of numbers. It is thus an almost impossible language in which to write a computer program, so higher-level languages are used that contain actual instructions expressed in words. High-level languages are translated into machine code by a compiler program. An assembly language is a low-level language that contains the same instructions as machine code, but expressed in names rather than numbers. Assembly language is translated into machine code by an assembler program.

Mach number The ratio of the speed of a body to the speed of sound at sea level (740 miles per hour). A body traveling at a speed in excess of Mach 1 is moving at supersonic speed. Mach 2 is 1,480 miles per hour, and so on. The term honors Austrian physicist and philosopher Ernst Mach who proposed the system.

macro A macro-instruction; a sequence of commonly used computer instructions contained in a short program that can be called up and put into operation with a single command. The use of macros can save a great deal of time in programming at all levels and in performing any repetitive computer task.

Magellanic Clouds Two dwarf galaxies, satellites of the Milky Way, about 180,000 light-years away containing a few billion stars.

magnadur A material made from ferromagnetic compounds, used to make permanent magnets, for example, for use in the focusing sections of a television tube.

magnet An iron-containing body with domains sufficiently well aligned to produce an external magnetic field and to experience a turning force (torque) when placed in another magnetic field.

magnetic field A space surrounding a permanent magnet or an electric conductor carrying a current in which a magnetic force can be detected by its effect on iron-containing bodies or on charged particles. Magnetism can be induced into a suitable body by bringing it into a magnetic field, and an electric current will be induced into any electrical conductor that is moved in a magnetic field.

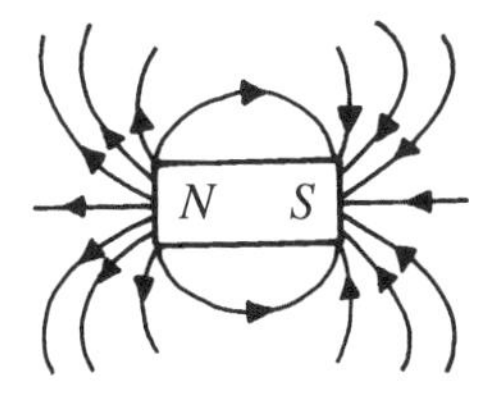

Magnetic field

magnetic flux *See* flux.

magnetic flux density The magnetic flux passing through a unit area of a magnetic field normal to (at right angles to) the magnetic force.

magnetic induction Magnetization induced in a magnetic material by being placed in a magnetic field. *See* magnetic flux density.

magnetic moment The relationship between a magnetic field and the turning force (torque) it exerts on a magnetic body or a charged body or electric current.

magnetic permeability The readiness to which a material can be magnetized by a given magnetic field. Permeability is proportional to the strength of the induced magnetism (flux density) divided by the intensity of the field.

magnetic resonance A phenomenon based on the fact that the inner core (nuclei) of atoms spins constantly. Protons carry a positive charge, so atomic nuclei, in spinning, create a tiny magnetic field. Hydrogen atoms behave in this way. When such atoms are placed in a strong magnetic field, the two magnetic fields interact and all the atoms are forced into alignment in a particular direction. The spin frequency of these atomic nuclei is known, and when another set of radio signals (an electromagnetic field), oscillating at this frequency, is briefly applied at an angle to the main field, the axes of these spinning nuclei are turned through an angle. The return movement to their former position, after the applied pulse ceases, causes these atomic magnets to generate tiny electromagnetic signals (radio waves), and these can be detected and the position of their origin accurately calculated and recorded.

magnetic resonance imaging (MRI) A method of obtaining detailed visual information about the internal structure of an object, such as the human body, using the principle of magnetic resonance.

magnetism A class of phenomena in which a field of force is caused by the alignment of the magnetic moments of atoms in the same direction (domains) – ferromagnetism – or by a moving electric charge – electromagnetism. Diamagnetism is the phenomenon caused by the orbital motion of electrons in the atoms of any material. Paramagnetism is due to the alignment of unpaired spins of electrons in atoms of a material. All materials thus respond to some extent to a magnetic field, but only ferromagnetic materials show a readily apparent response.

magnetohydrodynamics A propulsion system for sea vehicles in which an electric current between immersed electrodes generates a magnetic field that is repulsed by the powerful field of superconducting magnets. The conducting salt water is driven out of the open back of the vehicle, creating a forward thrust.

magnetomotive force The work done when a magnetic pole moves around a magnetic circuit. It is the magnetic equivalent of electromotive force.

magnetosphere The area around the Earth that experiences the magnetic effects caused by the Earth's magnetic field.

magnetostriction The change in the dimensions of an iron-containing body when it is subjected to a magnetic field. The direction of the change depends on the direction of the applied magnetic field. Magnetostriction is applied in various transducers, especially those producing high-frequency sound (ultrasound). Such a device consists simply of a coil of wire wrapped around a rod of the ferromagnetic material.

magnification The comparison of the size of an object to the size of the image produced by an optical system (lens, mirror, microscope, binoculars, telescope). Magnification may be linear or angular. Linear magnification is the ratio of the size of the image to that of the object. Angular magnification is the ratio of the angles formed at the eye by the image and the object.

magnifying glass A thick convex lens used to produce an erect magnified image of an object. The object is placed between the lens and its focal point and is viewed through the lens.

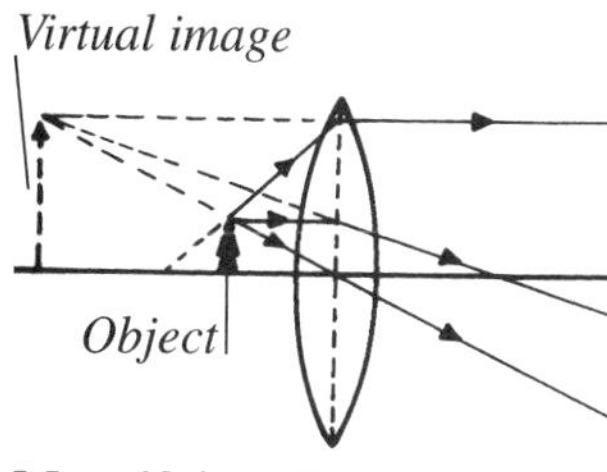

Magnifying glass

malleable Describing a material that can be shaped by hammering or rolling; it does not fracture easily.

Maltese cross tube A cathode-ray tube in which a piece of metal in the shape of a Maltese cross is connected to the anode. When a beam of electrons is emitted by the cathode, a sharp shadow of the Maltese cross is formed on the fluorescent screen, showing that electrons travel in straight lines.

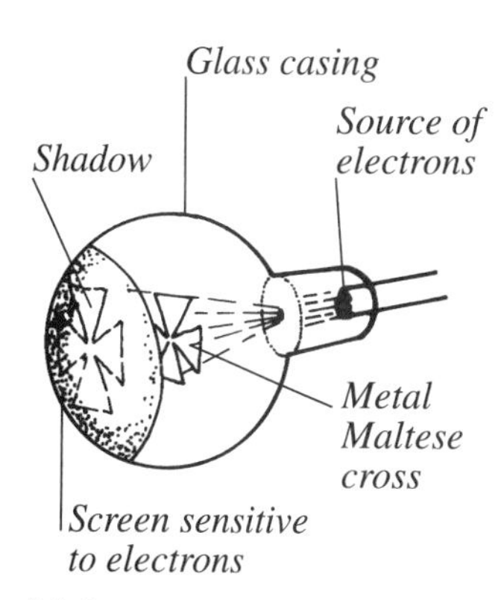

Maltese cross tube

manometer An instrument for measuring pressure, especially in gases. A manometer consists of a U-shaped glass tube containing a liquid, such as water or mercury, and a scale placed behind the upper level of the liquid in one arm. If a single pressure is to be measured, the source of pressure is connected to the top of one arm, so that a rise of pressure forces the liquid down in that arm and up in the other. If two pressures are to be compared, connections are made to both arms.

mantle, Earth The layer within the Earth that lies between 1,823 miles (2,900 km) below the surface and the crust.

maser Acronym for *m*icrowave *a*mplification by *s*timulated *e*mission of *r*adiation. It is similar in operation to the laser but produces longer-wavelength radiation.

mass The measure of a body's resistance to acceleration. Mass, unlike weight, does not change when subjected to different gravitational forces.

mass number The total number of protons and neutrons (nucleons) in the nucleus of an atom.

mass spectrometer An instrument used to measure relative atomic masses accurately, taking into account the analysis of different isotopes. In the spectrometer, ions of the gaseous sample are deflected, and the extent of their deflection indicates the presence and quantity of particles with different ratios of charge to mass.

matrix (1) (base of composite) The continuous component of a composite material in which other, tougher materials are embedded.

(2) In computing, a rectangular array of addresses consisting of horizontal rows and vertical columns. The rows are numbered from top to bottom and the columns form left to right. Any address can thus be identified by referring to the appropriate row and column. (3) In commerce, mathematics, statistics, engineering, economics, psychology, and many other disciplines, a spreadsheet laid out as in (2), used as computational techniques to solve a wide variety of problems.

Maxwell's equations Laws of theoretical physics governing the behavior of electromagnetic waves.

mean free path The average distance traveled by molecules between successive collisions. In gases this is inversely proportional to the pressure.

mechanical advantage A measure of the effectiveness of simple machines, such as the lever. Mechanical advantage is the ratio of the force exerted by the machine to the force applied to it. In general, an increase in the output force will be associated with a decrease in the distance through which it operates compared with the distance through which the input force operates.

mechanical energy The energy possessed by a body because of its motion, position, or state of tension or compression.

mechanics The study of the effect of forces on bodies. Dynamics, kinematics, and statics form part of mechanics.

mega- (M) A prefix denoting a multiple of 1 million. One megabyte (1 Mbyte) is a million bytes. One megahertz (1 Mhz) is a million cycles per second.

melting point The temperature at which a pure solid changes state from solid to liquid at a certain pressure. *See* freezing.

membrane A thin layer of a substance perforated by tiny uniform channels. It will allow small particles to pass through but stop larger particles.

memory In computing, a device or medium capable of retaining information in such a way that it can be recalled as often as required. A volatile memory is a semiconductor memory whose contents are retained only as long as the power supply is maintained. A random access memory (RAM) is a volatile matrix memory in which data is stored at (written into) known addresses, from which they can be retrieved and left or written over as required. A read-only memory (ROM) is a nonvolatile memory that cannot, in normal practice, be written to, but can be read from whenever needed.

meniscus The curved surface of liquid in a narrow tube. The surface is concave for water and other liquids, which are attracted more by the walls of the tube than other liquid molecules. For mercury in a glass tube, the surface is convex, as the mercury molecules near the glass are attracted less by the glass than by the other mercury molecules.

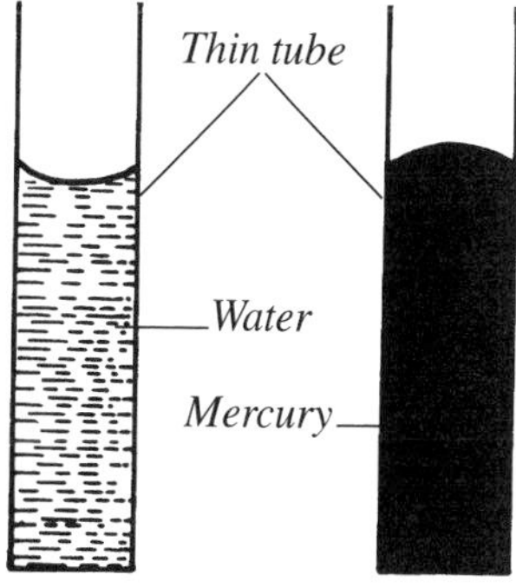

Meniscus

mercury vapor lamp A quartz tube in which an electrical discharge through ionized mercury vapor produces brilliant blue-green light and strong ultraviolet radiation.

meson An unstable medium-weight subatomic particle containing two quarks. Mesons are hadrons. There are several different mesons, which have different charges, positive, negative, and zero. Kaons, pions, and η mesons are some of the different types.

metallic salt A compound formed when one or more hydrogen atoms in an acid are replaced by a metal ion.

micro- (μ) A prefix denoting a submultiple of one-millionth of any particular unit. A microgram (μg) is a millionth of a gram. A microsecond (μs) is a millionth of a second.

microlithography The use of a very fine beam of laser light, electrons, or X rays to cut the microscopic circuit detail in the masks from which electronic microchips are made. Shadow-masking using these masks and light or X rays on silicon wafers coated with a light-sensitive resist allows the pattern of the circuit to be transferred to the wafer in terms of unaltered resist, which can then be removed with a solvent.

micrometer screw gage An instrument used to measure small distances accurately. The object to be measured is placed in the U-shaped gap, which is then tightened by a ratchet until it is gripped. The diameter of the object is read by combining readings on the barrel and shaft.

microprocessor A computer central processor unit in a single integrated circuit chip. A microprocessor forms the heart of a microcomputer but is used in many other applications, such as process controllers in domestic and industrial equipment, interactive toys, machine tool controllers, point-of-sale and other terminals, and car-fuel and engine-control systems. Microprocessors vary considerably, but all contain an arithmetic and logic unit (ALU), a control unit, a control memory, a bus control, an internal memory, and a working register.

The whole operation of the microprocessor is kept in synchronization by a clock that produces a square-wave output. The microprocessor also has its own software in the form of an instruction set for carrying out input and output operations, logic, arithmetic, data transfer, and so on.

microscope An instrument used to obtain magnified images of small objects. The magnifying glass is a simple microscope. The compound microscope consists of two convex lenses with short focal length, the objective and eyepiece lenses, fitted at opposite ends of a tube. A virtual magnified inverted image of the object is produced. The magnification produced is the product of the magnification of the objective and eyepiece lenses.

Eye
Ray path
Object
Illuminating rays

Microscope

microwaves Electromagnetic radiation lying between radio waves and the infrared in the spectrum. Microwaves have wavelengths of a few centimeters to a few millimeters and frequencies of hundreds of millions of cycles per second (hundreds of megahertz) to tens of thousands of megahertz. They are produced by oscillators, such as klystrons and magnetrons, and are conducted along wave guides or coaxial cables. They can be transmitted in the same way as other radio waves and have the advantage that antennas are appropriately small. They are usually transmitted in beams, using small dipole aerials at the focus of parabolic dish reflectors. Because of their high frequency, microwave bands provide a wide range of usable channels. They are used in telecommunications, radar, heating, and cooking.

mild steel (low carbon steel) It contains between 0.04 and 0.25% carbon. It is soft and is used for car bodies, ships, etc.

milli- (m) A prefix denoting a submultiple of one-thousandth of any unit. One milligram (1 mg) is a thousandth of a gram. One millisecond (1 ms) is a thousandth of a second.

mips Acronym for *m*illions of *i*nstructions *p*er *s*econd. This is a measure of the speed and power of a computer.

mixture A system that consists of two or more substances (solid, liquid, or gas) present in any proportion in a container. There is no chemical bonding between substances. A mixture can be separated using physical methods. The formation of a mixture does not involve a change in temperature.

modem A device that enables computers to communicate with each other along ordinary telephone lines.

moderator A material used in a nuclear reactor to slow fast neutrons so that they will undergo fission reactions. Graphite and heavy water are used as moderators.

modular programing The normal method of proceeding, in which programs are written in sections (modules), each of reasonable and manageable size, and each performing a particular, well-defined function. Modules, once written and tested, can be exported to other programs.

modulation The process of superimposing a signal carrying information onto another wave (the carrier wave) or onto an electron beam. The superimposition may involve changing the amplitude, the frequency, the phase, or the continuity of the carrier wave. The carrier wave will usually have a much higher frequency than the information signal, as in the case of amplitude modulation (AM) radio, in which the carrier wave may be of a frequency of many megahertz, while the information is of audio frequencies (about 20 to 20,000 Hz). In frequency modulation (FM), the amplitude remains constant, while the frequency is changed at a rate corresponding to the changes in the information. In phase modulation, periodic alterations occur in the timing of the start of each cycle of the carrier wave (phase), in accordance with the information frequencies. In pulse code modulation (PCM), the carrier is switched on and off in pulses to form a binary code conveying the information.

Mohs' scale A scale that measures the hardness of minerals by their ability to scratch one another. A mineral is given a number on Mohs' scale according to its ability to scratch one of the reference materials. In order of increasing hardness they are: 1, talc; 2, gypsum; 3, calcite; 4, fluorite; 5, apatite; 6, feldspar; 7, quartz; 8, topaz; 9, corundum; 10, diamond.

molality (m) Concentration of solution giving the number of moles of solute dissolved in 1 kg of solvent.

molar gas constant The universal gas constant (R). It is used in the gas equation $PV = nRT$.

molarity (M) Concentration of solution giving the number of moles of solute dissolved in 1 dm^3 of solution.

mole The amount of a substance that contains the same number of entities (atoms, molecules, ions, any group of particles, but the type must be specified) as there are atoms in 0.012 kg of the carbon-12 isotope.

The actual number is known as the Avogadro number; its value is 6.023×10^{23}.

molecular energy The kinetic energy of a molecule in a gas. At any temperature the molecules in a gas possess a range of kinetic energies.

molecule The smallest part of an element or chemical compound that can exist independently with all the properties of the element or compound. It is made up of one or more atoms bonded in a fixed whole number ratio.

moment The turning effect of a force around a point.

moment of a force or **torque** The product of a force and the perpendicular distance from the line of action of the force to the turning point.

moment of inertia Taken over the whole of a body, it is the sum of the products of the body's mass and the square of the distance of that mass from the body's axis of rotation. It gives the tendency of the body to resist angular acceleration and can be expressed in the form mk^2, where m is the mass of the body and k is the radius of gyration about the axis.

momentum A body's momentum is the product of its mass and velocity.

monochromatic light Electromagnetic radiation having only one wavelength and, hence, one color.

MOSFET Acronym for *m*etal *o*xide *s*emiconductor *f*ield *e*ffect *t*ransistor, a type of transistor fabrication suited for large-scale chip integration, with a low power consumption and high switching speed, compared to earlier transistors. A high proportion of computer chips are currently of MOSFET construction.

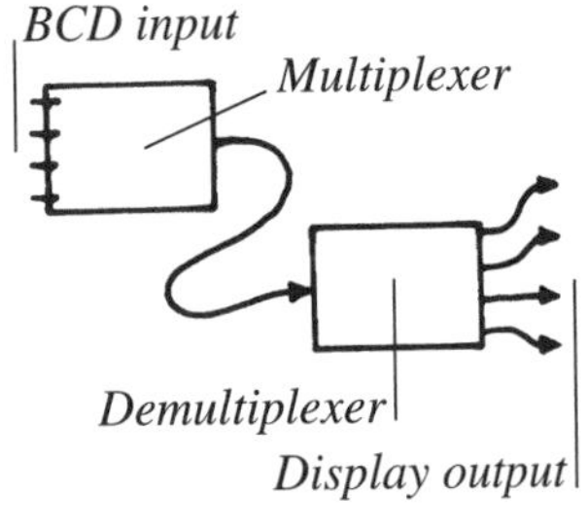

Multiplexing

multiplexing The combining of a number of different signals from separate sources in a common channel of communication so that they can be transmitted together and later separated. This can be done in a variety of different ways. In time-division multiplexing, for instance, a sequence of consecutive very short time slots is used, and these are allocated, one after the other, to the different signals. Thus signals A, B, C, and D might each be allocated one-thousandth of a second each in which to transmit. After four-thousandths of a second, the cycle is repeated, and so on. When the signals are separated, the gaps are inappreciable.

muon A subatomic particle with low mass, a lepton. The muon has a negative charge, a mass 207 times that of the electron, and a very brief life.

mutual inductance The generation of an e.m.f. in one system of conductors as a result of changing magnetic flux in an adjacent system of conductors; it forms the basis of a transformer.

nano- (n) A prefix denoting a submultiple of one-billionth of any unit. One nanosecond (1 ns) is one-billionth of a second.

nanotechnology (1) The physical manipulation of matter at molecular level by effecting actual movement, using the tip of a scanning tunnel microscope.
(2) The production of ultraminiaturized microscopic devices or machines that can perform useful functions in areas otherwise inaccessible to such manipulations.

nebula A celestial object composed of gas and dust that appears as a hazy smudge of light when viewed through an optical telescope.

neutral (1) (*chemistry*) A solution whose pH is 7. It is neither acidic nor alkaline.
(2) (*physics*) A conductor having neither negative nor positive electric charge; it is at Earth potential.

neutralization The reaction of an acid and a base, forming a salt and water. The properties of acids and bases disappear when the reaction is complete, at the end point. The resulting solution is neutral.

neutrino A subatomic particle with zero charge and zero mass; a lepton.

neutron One of the two major components of the atomic nucleus, the other being the proton. Only hydrogen-1 has no neutron in the nucleus. The neutron has a mass very close to that of the proton but has no electric charge. Protons are positively charged, and since like charges repel each other, two or more protons could not exist together within the dimensions of atom nuclei were it not for the presence of neutrons. Neutrons weaken the electrostatic repulsion without weakening the forces that bind nuclear particles together. In the lighter elements the number of neutrons and proton is about equal; in the heavier elements the number of neutrons predominates. Isotopes are different varieties of the same element having the same number of protons but different numbers of neutrons. Free neutrons emitted from nuclei are needed to sustain nuclear fission chain reactions.

neutron mass 1.675×10^{-27} kg.

neutron number The number of neutrons in the nucleus of an atom. All isotopes of an element have the same atomic number but different neutron numbers.

neutron star An astronomical body formed after a supernova explosion. Its diameter is a few kilometers but its mass is approximately equal to that of the Sun because it is extremely dense. As a result of its very high density, protons and electrons combine to form neutrons within it.

newton The SI unit of force. It is the force that will give a 1-kilogram mass an acceleration of 1 m sec^{-2}.

Newton's disk A disk painted (in any order) with the seven spectrum colors of white light. This disk appears white when spinning at high speed, demonstrating that white light is composed of these colors.

Newton's laws of motion (1) A body will continue at rest or in a state of constant velocity unless acted on by an external force.
(2) The rate of change of a body's momentum is proportional to the force applied and is in the direction of the force.
(3) If two bodies interact, the first body exerts the same force on the second as the second does on the first, but in an opposite direction, action and reaction are equal and opposite.

Newton's rings Concentric rings (dark and light if the light used is monochromatic, colored if it is white) seen when a plano-convex lens is placed on a flat glass plate and illuminated from above. The rings are caused by interference between light reflected from the glass plate and light reflected from the curved surface of the lens.

Nicol prism Used to obtain plane-polarized light. It consists of two crystals of calcite cemented together with Canada balsam. When an ordinary light ray strikes the prism, it is split two ways (double refraction). The ordinary ray is reflected from the balsam layer (total internal refraction). The extraordinary ray passes straight through the prism, emerging plane-polarized.

nitrogen cycle The circulation of nitrogen in nature between organisms and their environment. It involves the fixing of gaseous nitrogen for use by higher organisms, the absorption of nitrogenous substances from the soil and water, and the return there by decomposition.

nitrogen fixation The conversion of atmospheric nitrogen into nitrogenous substances. This occurs naturally by the action of certain soil bacteria on the nitrogen in ammonia, and the conversion of atmospheric nitrogen to its oxides by lightning. The Haber process is an industrial process for the fixation of nitrogen.

noble gases (inert gases or rare gases) Helium, neon, argon, krypton, xenon, radon. They occupy the group furthest right on the periodic table, as they all have complete octets of electrons in their outer shells. They are chemically very inert monatomic gases. They are used in gas discharge tubes.

node A point on a standing wave where there is zero displacement. The distance between two nodes is half the wave's wavelength.

noise Any unwanted signal or information added to a communication channel in the course of its use. Noise need not be, indeed, seldom is, acoustic in nature, although it may be audible if the information leaves the communication channel in audible form. It is usually electrical in nature and is quantified in terms of the signal-to-noise ratio. Noise may, for instance, be visible on a VDU as light flashes. Noise can degrade the quality of an information channel by producing spurious signals and by rendering voltage logic levels ambiguous. A TV screen displays considerable noise after TV transmission has ceased. Frequency modulation (FM) transmission is inherently less sensitive to noise than amplitude modulation (AM) transmission.

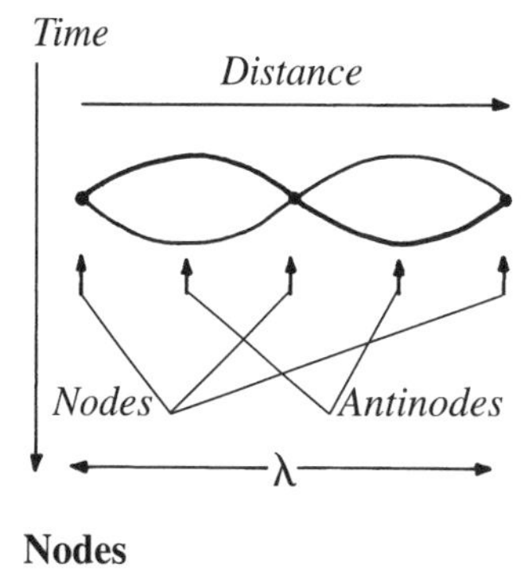

Nodes

nonmetal An element that is not a metal. Nonmetals readily form negative ions and are poor conductors of heat and electricity; their oxides are acidic.

normal In mathematics and physics, a term meaning perpendicular to, or perpendicular to the tangent of a curve at a particular point.

nuclear binding energy The difference between the mass of an atom and the sum of the masses of the protons, neutrons, and electrons of which it is constituted. This mass difference, expressed in energetic terms, is the nuclear binding energy that holds the atom together.

nuclear bomb A bomb where the explosive power is caused by uncontrolled nuclear fission or fusion.

nuclear energy (atomic energy) The energy released, principally in the form of heat, light, and radiation, as a result of changes in the nuclei of atoms. Nuclear energy is released in certain natural processes, such as the spontaneous decay of naturally occurring radioactive substances, and the nuclear reactions that power the Sun and other stars. It is also released in human-made devices such as nuclear reactors and nuclear weapons.

nuclear fission *See* fission.

nuclear fusion *See* fusion.

nuclear magnetic resonance The nuclei of some isotopes will behave like magnets and align themselves in a strong magnetic field. If they are exposed to radio-frequency radiation, this alignment can be changed. The frequency at which this change occurs is the resonant frequency and depends on the type of nucleus. This principle is used to investigate structures and for magnetic resonance imaging of the human body.

nuclear waste Waste can be a mixture of isotopes with very short and very long half-lives. Waste is kept in storage tanks to allow high-level wastes to decay to less active or stable isotopes.

nucleon A proton or neutron.

nucleon number or **mass number** The number of nucleons (protons and neutrons) in the nucleus of an atom.

nucleus The small (about 10^{-14} m diameter) core of an atom. All nuclei contain the positively charged proton, and all but hydrogen contain the zero-charged neutron. The sum of protons and neutrons is the atom's mass number (or nucleon number). The nucleus is surrounded by a cloud of electrons whose number is equal to the number of protons in the nucleus.

nuclide A particular isotope of an element, identified by the number of protons and neutrons in the nucleus.

objective lens In an optical instrument (microscope, telescope) it is the lens nearest to the object.

object-oriented programing (OOP) A type of computer programing in which the formerly conventional ideas of the distinction between data and procedures are set aside. They are replaced by ideas of objects and messages. Objects are collections of information, each with a description of its particular manipulation; messages are specifications of what is to be done with objects. OOP languages include Smalltalk, C and C++, and Java, and these are rapidly replacing the previously more widely used languages.

ocean thermal energy conversion (OTEC) OTEC systems can generate about one megawatt of electricity by using the thermal gradient of the ocean, which, in some areas, can be of the order of 20°C over a depth of several hundred feet. The energy involved in this temperature difference is converted, following thermodynamic principles (*see* heat pump), to mechanical energy, which is then converted to electricity.

Ohm's law The ratio of the current (I) in a metallic conductor at constant temperature to the potential difference (V) between its ends is constant. This constant is the resistance of the conductor; thus $R = V/I$.

open-hearth furnace An industrial process for steel manufacture. Pig iron, scrap steel, and limestone are heated to very high (1,750°C) temperatures to remove impurities that enter the slag that forms on the liquid metal surface and is removed. Steel with specific qualities is obtained by adding suitable elements, such as carbon and manganese.

operand A number, quantity, or function on which a mathematical or logical operation is performed.

operating system The software that controls the basic running of a computer.

operational amplifier (op-amp) A very high-voltage gain, directly coupled, usually integrated-circuit amplifier with a large amount of negative feedback from output to input, a very high-input impedance, and a low-output impedance. It has two inputs and a single output, which becomes positive when the positive input is more positive than the negative input, and vice versa. Gain is stabilized by the negative feedback, and the absence of capacitors in the coupling between the stages allows amplification at frequencies down to DC. Operational amplifiers have been used extensively in analog computing (and in instrumentation) and are still widely used as conveniently packaged general-purpose voltage amplifiers.

optical activity The ability of different forms of the same compound, in solution, to rotate the plane of a beam of plane-polarized light, either clockwise or counterclockwise. The effect is due to the fact that some molecules exist in two forms, which are mirror images of each other but have an asymmetry, like that of the right and left hands, that prevents them from being superimposed on each other. These forms are called optical isomers or enantiomers. Dextrorotary forms may have subtly different properties from laevorotary forms, in that they may interact differently with other asymmetric molecules. Some drugs, for instance, are more active in one form than in the other.

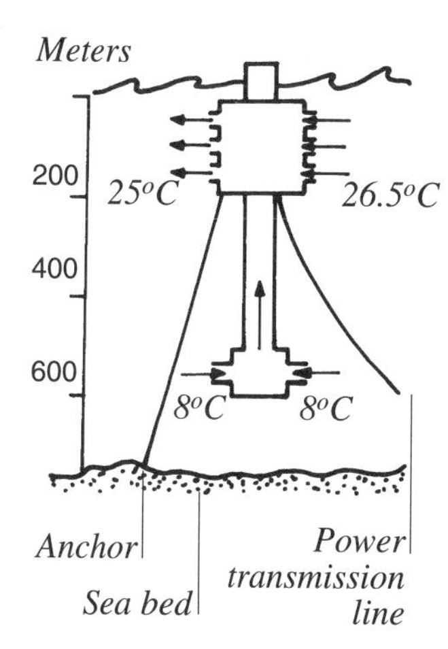

Ocean thermal energy conversion

optical character recognition (OCR) A method of converting printed text to digital code so that it can be stored, rapidly accessed, and used by computers. Text must first be scanned to produce a computer-storable image of the page. The position of the pixels of each letter is then compared with those of a stored vocabulary of letters of the same typeface until a match is found. The event is used to generate the ASCII code for the letter concerned. Spaces generate the code for a space. OCR software can be "taught" to recognize unusual typefaces, hand-printed letters, or even normal handwriting. OCR is currently being used to convert all printed material worth preserving to machine-usable code.

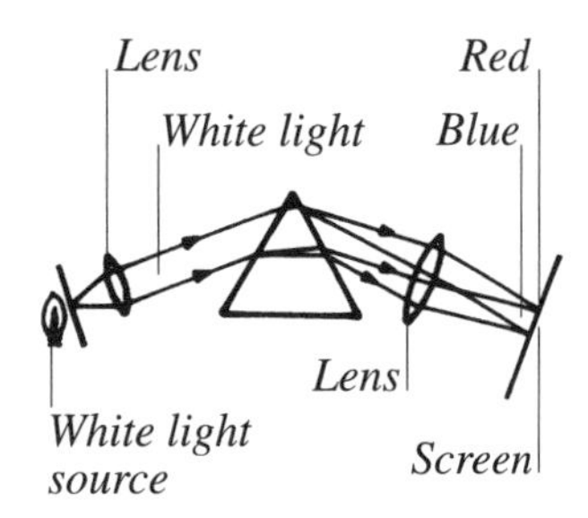

Optical dispersion

optical dispersion A beam of light of mixed colors is dispersed into its component spectrum colors by passing through a different medium. The different components of the spectrum travel at different speeds and are hence refracted at different angles.

optical fibers Very fine filaments of glass in a protective sheath along which light can be transmitted. It can be modulated with other signals to provide communications.

optoelectronics The interface between electronics and associated light signals conveying information. Optoelectronics is concerned, for instance, with the conversion of information in the form of electrical signals into a form that can be transmitted in a laser beam along a glass fiber optic channel, and the reconversion, at the other end, into electronic form. It is also concerned with the possibilities of implementing logic by means of lasers and ultra-rapid light switches. Liquid crystal displays are another aspect of optoelectronics.

orbit The path of a heavenly body (or artificial satellite) circling another as a result of the attraction of gravity.

order of reaction A first-order reaction is one in which there is a spontaneous decomposition of one molecule; one that takes place with two molecules is a second order reaction, and so on.

ore A mineral from which a metal or nonmetal may be profitably extracted.

organic compounds Compounds containing carbon but not carbonates or carbon dioxide.

oscillation A rapid backward-and-forward repetitive motion in a fluid or solid that behaves elastically.

oscillator An electronic circuit that converts direct current electricity into an alternating current at a relatively fixed or variable frequency. There are many different kinds of oscillators, and they are extensively used in electronics. Many are amplifiers with positive feedback from output to input. Various elements, especially inductance and capacitance in the amplifier, determine the natural resonant frequency of the circuit. Many oscillators employ a tuned circuit consisting of a capacitance in parallel with an inductance in the feedback loop, and these are inherently stable. Even more stable are quartz crystal oscillators. Relaxation oscillators contain a capacitor that charges at a certain rate until the voltage across it reaches the level at which a device, such as a neon lamp, connected in parallel with the capacitor, begins to conduct. As it does, the capacitor is short-circuited, and the voltage across it drops well below the striking voltage of the lamp, which goes out. The cycle is then repeated.

oscilloscope A scientific instrument consisting of a cathode-ray tube and associated electronic circuitry that allows rapidly varying voltages to be accurately represented in graphical form. The oscilloscope can display the details of the waveform of periodic voltages of all frequencies, can measure frequencies (*see* Lissajous' figures), and can compare the simultaneously displayed waveforms of a signal before and after its passage through an electronic circuit or filter network. Software can be used to convert a personal computer into an oscilloscope.

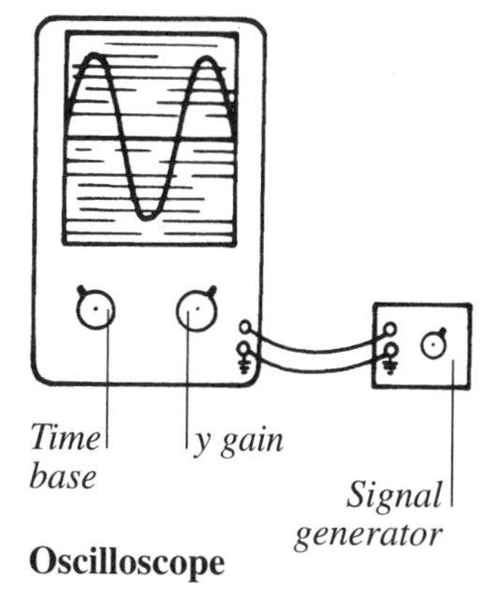

Oscilloscope

Otto cycle An engine cycle for a four-stroke gasoline engine in which the processes all take place at constant volume with two revolutions of the crankshaft.

overtone *See* harmonic.

oxidation A substance is oxidized if it gains oxygen, loses hydrogen, or loses electrons.

oxidation state (oxidation number) This gives an indication of the electron control that an atom has in a compound compared with that which it has in a pure element. It is in two parts: one is the sign – if control has increased, it is negative, if it has decreased, positive. The other part is the value, which gives the number of electrons over which control has changed. In oxidation there is an increase in the number.

oxide A compound of oxygen and one other element. There are four groups: acidic oxides (oxides of nonmetals), basic oxides (oxides of metals), neutral oxides (carbon monoxide), and amphoteric oxides (having basic and acidic properties, aluminum and zinc).

ozone One of the two allotropes of oxygen, existing as O_3. It is a bluish gas with a penetrating smell. It is a very strong oxidizing agent.

ozone layer or **ozonosphere** A layer of ozone in the Earth's stratosphere that absorbs much of the ultraviolet radiation from the Sun, thus shielding the Earth from its harmful effects. It is at an altitude of between 12 and 50 km.

packet switching A widely used method of digital communication between multiple points. Messages are divided into segments of fixed size, and each segment carries a code identifying the addressee. Packets are sent serially, without regard to the addressees, so as to fill up the communication time and space available and avoid wastage. The identifying code ensures that addressees receive only the appropriate packets. These are then reassembled in order into complete messages.

packing fraction A measure of the stability of a nuclide. Its mass (*M*) in atomic mass units is slightly different from the mass number (*A*). Packing fraction = $(M–A)/A$. Mass defect is $M–A$.

Parabolic reflector

parabolic reflector A concave reflector, a section across which is a parabola (the conic section formed by the intersection between a cone and a plane parallel to its side). A parallel beam of light incident on a parabolic reflector is converged to its focal point. If a light source is placed at the focal point of a parabolic reflector, it produces a parallel beam of light.

parallax The apparent change in an object's position against a more distant background when it is seen from different places.

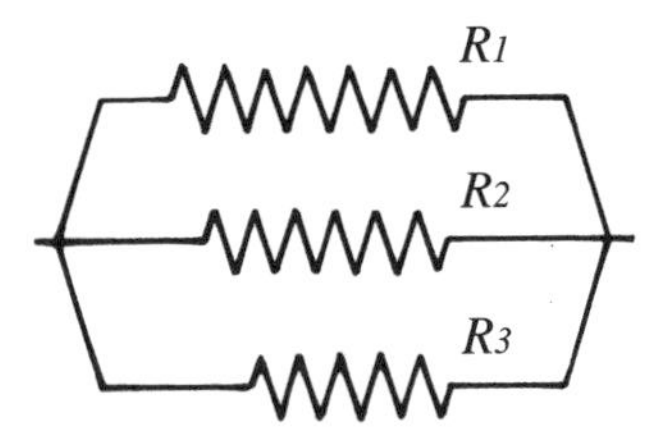

Parallel resistors

parallel In plane Euclidean geometry, two straight lines are considered to be parallel to each other if they do not meet. They are separated by an equal distance at every point. If two or more electrical components are connected between the same points in a circuit, they are said to be connected in parallel.

paramagnetism Magnetism caused by unpaired electrons. An applied magnetic field will tend to align the magnetic moments of the atoms, but this will disappear in the absence of the field.

parameter An arbitrary quantity in a mathematic function whose value is selected as required. Also, a variable in terms of which other interrelated variables can be expressed and can be regarded as being dependent on the parameter. An auxiliary variable whose functions give the coordinates of a curve or a surface.

parametric amplifier An amplifier for very high frequencies or microwave frequencies, consisting of a variable inductor or capacitor whose reactance is altered at an even higher frequency by another pump signal. Varactor diodes can behave in this way and are commonly used. The resultant mixing of frequencies produces additional signals at beat frequencies, and these can be at a higher power than the input frequency, gaining energy from the pump signal, thus effecting amplification. Parametric amplifiers contribute a very low noise level to the signal and are often used as the input stages of microwave receivers.

parity bit A binary digit added to a byte as a means of checking that the byte has not been changed in transmission.

partial tone The same as harmonic.

particle (1) In classical physics, a body of small but finite mass but of no appreciable size. As a result, although a particle has inertia and is susceptible to gravitation, any force acting on it can cause only displacement and cannot cause rotational acceleration.
(2) Any elementary component of matter, especially at a subatomic level.
(3) A solid or a liquid in a finely divided state but still consisting of a large number of molecules. A collection of dry solid particles is called a powder. Particles suspended in a fluid produce a dispersoid or hydrosol. A suspension of particles, either solid or liquid, in a gas is called an aerosol.

particle accelerator A device such as the linear accelerator (charged particles are accelerated in a straight line by an electric field), cyclotron, or synchrotron (electrical and magnetic fields combine to accelerate charged particles in a circular path) that creates high-energy particles that are used in collisions to produce other subatomic particles in the investigation of fundamental particles.

Pascal A structured programming language, once widely used but now being superseded by object-oriented programming languages. Pascal has been much used for teaching computer science.

Paschen series A series of lines in the infrared part of the hydrogen emission spectrum.

passive star network A computer communications network arrangement in which peripheral terminals are connected by radiating links to a central node, which simply connects them to each other without any processing of the information.

patina A compound coating a metal or mineral in a thin, often multicolored film. It is caused by the reaction of the metal or mineral with the atmosphere.

peak value of alternating current The maximum positive or negative value of an alternating current.

peak value of alternating e.m.f. The maximum positive or negative value of an alternating source of e.m.f.

peek To use a high-level language to examine the contents of an address in a computer memory. To *poke* is to use such a language to insert a datum into a memory address or to modify its contents.

Peltier effect The rise or fall in temperature that occurs at the junction of dissimilar metals when an electric current flows through it. The direction of temperature change depends on the direction of current flow. The phenomenon can be put to practical use if semiconductor junctions are employed. The effect was first noted in 1834 by the French physicist J. C. A. Peltier.

pendulum An elongated body free to swing on a pivot, under the influence of gravity, about a horizontal axis. The lower end of the pendulum describes simple harmonic motion. The time for one complete excursion of the swing (the period) is independent of the mass or weight of the pendulum and is affected only by its length and by the local value of the gravitational constant g. The period is also almost independent of the amplitude of the swing. The time for a complete swing both ways is given by $T = 2\pi\ (l/g)$, where l is the length of the pendulum.

period The time taken for a complete cycle of a process that is repeating regularly.

periodic function A function that regularly repeats itself. For instance, the sine wave graph of $y = \sin x$ repeats itself every 360° and is a periodic function.

periodic motion Any movement that precisely repeats itself, taking equal intervals of time to do so. The swing of a pendulum is an example of periodic motion, as are the mass movements of electrons in alternating current. A graph of the voltage changes in an AC supply is a representation of periodic motion.

periodic table A table that classifies elements into periods (the horizontal rows) and groups (the columns). The elements are arranged according to increasing atomic number across each period and down each group. This classification shows the similarities and differences between elements and, knowing an element's atomic number, its properties can be predicted. The number of electrons in the outer shell is the same for each member of a group.

periscope An optical instrument used to look over or around an obstacle.

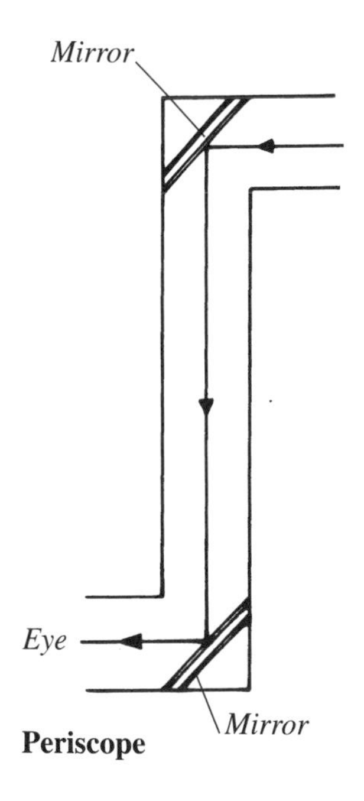

Periscope

permeability 1. (*materials science*) The extent to which a material allows the passage of a fluid through itself.
2. (*electrical science*) The magnetization in a material that is excited by a source of magnetomotive force. The absolute permeability is the ratio of the flux density produced to the magnetic field strength that produced it.

permeability of free space (magnetic field constant) It is used in calculations of relative permeability of a material. Its value is $4\pi \times 10^{-7}$ henry meter^{-1}.

permeable rock Rock that allows the passage of a fluid (water, oil, gas) through itself.

permittivity The electric flux density produced in a material that is excited by a source of e.m.f. The absolute permittivity is the ratio of electric flux density produced to the electric field strength.

permittivity of free space (electric constant, electric field constant) It is used in calculations of relative permittivity of a material. Its value is 8.854×10^{-12} farad per meter.

permutation One of all the possible arrangements of a set of numbers. There are $n!$ (factorial n) permutations of n numbers, taken all at a time. Factorial n means the product of all the integers up to and including n. Thus factorial 6 is $1 \times 2 \times 3 \times 4 \times 5 \times 6$, i.e., 720, so there are 720 permutations of all of any 6 numbers.

Perrin tube A cathode-ray tube used to demonstrate that a beam of cathode rays has a negative charge.

pesticide A chemical used to kill pests, particularly insects and rodents.

Pfund series A series of lines in the far infrared part of the hydrogen emission spectrum.

pH A scale that gives a measure of the acidity of an aqueous solution. The concentration of hydrogen ions is used in the calculations, and the pH value of a solution is given as ($\log^{10}(1/H^+)$), where H^+ is the concentration of hydrogen ions. A neutral solution has a pH of 7, while an acidic solution has a lower value, and an alkaline solution a higher value.

phase The point in a cycle of any periodic function or periodic quantity reached at a specific time. Since a periodic function can be said to repeat itself in 360°, phase can be expressed as an angle. Two or more periodic quantities can be said to be "in phase" or "out of phase" with each other, and this difference can be quantified in degrees. Equal amplitude waves 180° out of phase with each other will cancel each other.

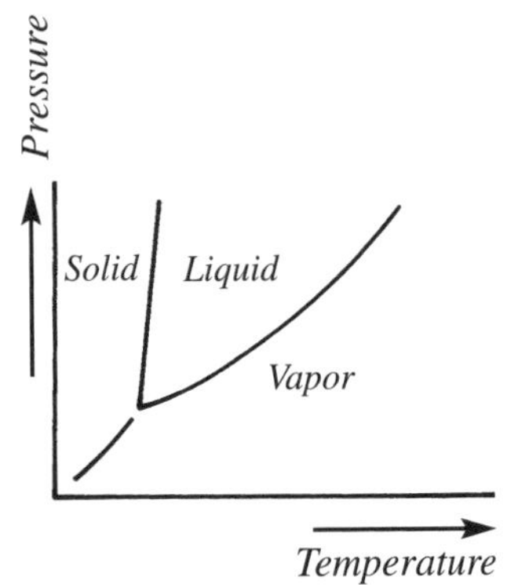

Phase diagram

phase diagram A diagram showing the change between states for a substance at different conditions of pressure and temperature.

phase-locked loop An electronic circuit containing a variably alternating-current generator (voltage-controlled oscillator) that is locked into phase with an incoming signal. The output is passed through a low-pass filter that removes all the alternating current, leaving the direct current signal. The effect of this is to remove noise from the signal and effectively regenerate an almost perfect replica of the signal. Should the signal vary in frequency, the phase-locked loop circuit will follow it and maintain the signal, which might otherwise be lost. The phase-locked loop principle is widely used in television and radio receivers, and the circuit is commonly fabricated on a single silicon chip as an integrated circuit.

phenolphthalein An acid-base indicator used in titrations of a weak acid with a strong alkali. It is colorless in acid solution and red when the solution is alkaline.

photocatalysis The speeding up of a chemical reaction by light.

photochemical reaction A photochemical reaction that requires light of a particular wavelength to initiate the reaction.

photoconductive cell A passive crystalline semiconductor device that changes its electrical resistance in proportion to the intensity of light falling on it. If the cell is included in a series circuit with a source of electricity and a sensitive electric meter, it can be used to measure light intensity. If the meter is replaced by a semiconductor or other relay, large currents can be switched by a small change in light intensity.

photoelectric cell A device that generates a voltage when irradiated with light. Such devices can be connected in parallel series to produce practical amounts of electric power. A photoemissive cell consists of a cathode that emits electrons when irradiated and a positively charged anode to attract the electrons and thus cause a current to flow.

Photoelectric effect

photoelectric effect The release of electrons from the surface of a conductor when electromagnetic radiation of a particular wavelength falls on the surface of the conductor (usually visible or shorter wavelength).

photographic film Film coated with a light-sensitive emulsion of a silver salt that is activated by light. These activated grains are more easily reduced to black metallic silver during the developing process.

photometry The measurement of light intensity, its color, the rate of flow of light energy (flux), and its point brightness (luminance).

photomultiplier A highly sensitive instrument capable of detecting and measuring very low levels of light intensity. The photomultiplier consists of a cascaded series of photoemissive stages (*see* photoelectric cell) in which the output of one stage forms the input of the next so that the gain rapidly rises.

photon A packet of electromagnetic radiation that can be considered as either a particle or a wave.

photosphere The visible surface of the Sun.

photosynthesis This is an important photochemical reaction. It is the process by which green plants make carbohydrates using carbon dioxide and water. Oxygen is produced.

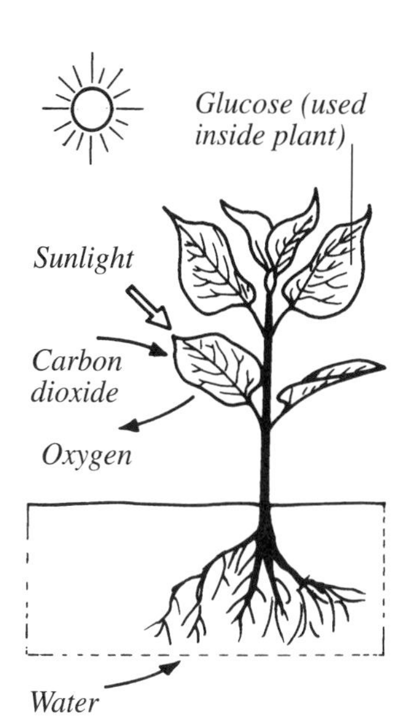

Photosynthesis

phototransistor A transistor with a light-sensitive base. A base current is created by the photoelectric effect, which is then amplified by the normal transistor effect.

physical properties The observed characteristics of a substance – boiling point, melting point, density, etc.

pi (π) The 16th letter of the Greek alphabet. It is important in mathematics, as it represents the ratio of the circumference of a circle to its diameter.

pico- (p) A prefix denoting a millionth of a millionth of any unit. Thus, a picosecond is 10^{-12} s.

piezoelectricity Voltages generated by certain dielectric crystals when subject to mechanical pressure or strain, and the deformation that occurs in the same materials when subject to an electric field. This double effect allows many practical applications as transducers. Piezoelectric crystals can resonate at a very precise frequency and are commonly incorporated into oscillators on this account. Every personal computer and every digital clock and wristwatch relies on piezoelectricity to stabilize its clock oscillator. Piezoelectricity is a feature of crystals of quartz, barium titanate, Rochelle salt, zincblende (ZnS), ethylene diamine tartrate, and other materials.

pig iron Impure iron produced from a blast furnace. The impurities it contains depend on the nature of the ore used (these can be carbon, silicon, manganese, sulfur, and phosphorus). The name is derived from the iron being cast into rough sand molds or "pigs."

Pipette

pinch effect The effect of the magnetic self-attraction occurring when an electric current flows. The inward pressure effect on metallic conductors can usually be neglected unless very large currents are flowing. The effect is used in one method of attempting to achieve a nuclear fusion reaction. A very large current flowing through a plasma (high-temperature gas with its electrons stripped off) will, by virtue of the pinch effect, raise the temperature and pressure even higher and will help to keep the area concerned clear of the walls of the containing equipment. Unfortunately, the pinch is unstable.

pion *See* meson.

pipette A glass tube that is used to measure and transfer a fixed volume of liquid. Pipettes are available in a range of volumes. Suction is applied to the top of a pipette to draw a liquid up, so that its meniscus is on the marked line on the pipette. When the suction is released, the liquid flows out of the pipette.

piston A round plate, attached to a rod, that moves within a cylinder in a pump or engine.

Pitot tube An open tube placed in an air stream. The side facing the flow receives the full pressure, while the other registers the static pressure. The dynamic air pressure is the difference between the two, and this is displayed on the air-speed indicator.

pivot or **fulcrum** The point of support about which a lever turns.

pixel or **picture element** The smallest element in a video display whose color and brightness can be controlled.

Planck's constant (h) The constant used in the calculation of the energy of quanta of electromagnetic radiation of a particular frequency. Its value is 6.626×10^{-34}.

Planck's law The energy of electromagnetic radiation exists as individual packets or quanta. Each has to be absorbed or radiated as a whole. The magnitude of the energy is proportional to the frequency. If E is the energy of the quanta and ν is the frequency, $E = h\nu$ where h is Planck's constant.

plane A flat surface.

plano-convex lens A lens with one flat surface and one convex surface.

plasma In physics, a gas consisting of roughly equal numbers of unassociated free positive ions and electrons. Plasmas readily conduct electricity and are influenced by electric and magnetic fields.

plasticity The property of being able to undergo a permanent change of shape as the result of the application of a stress greater than that needed to overcome its elastic limits.

plastic strain A permanent change to a material that has been stretched beyond its elastic limit.

plate tectonics The model used to explain various processes – mountain building, earthquakes, mid-ocean trenches – which suggests that the Earth's surface consists of large moving plates floating on a viscous mantle.

point charge A region of charge that has negligible area.

poke *See* peek.

polarity (of a molecule) The property of a molecule that has an asymmetrical distribution of electrons. It tends to have a positive and a negative end. Polar molecules act as solvents.

polarization Of a radiated wave, being oriented in a predictable direction perpendicular to the direction of propagation. In a plane-polarized radiation the waves are all confined to a particular plane. Ordinary radiation consists of transverse rays with vibrations in all possible directions perpendicular to the direct propagation of the wave.

Polaroid Trade name for a range of photographic and optical products. One of these is a transparent plastic sheet, composed of sheets of plastic containing many minute double-refracting crystals with parallel

Unpolarized waves

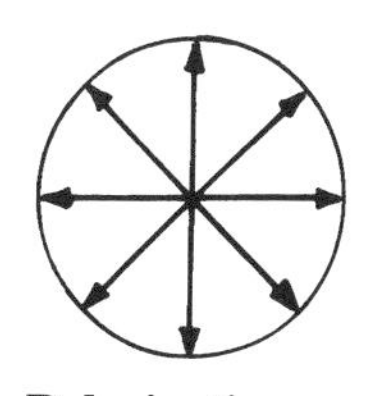

Partly polarized waves

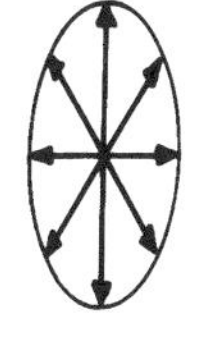

Polarization

axes, that produces plane-polarized light when ordinary light falls on it.

polymer A material containing very large molecules that are built up from a series of small basic units (monomers). It is often a term applied to plastics. There can be between hundreds and hundreds of thousands of basic units in a polymer.

polymerization The formation of a polymer from monomers. There are two types of polymerization techniques: addition and condensation.

porous Able to allow the passage of water, air, or other fluids.

position vector (directed line segment) It represents the displacement of a point from a reference point.

positron A subatomic particle with the same mass as the electron but positive charge (it is the antiparticle of the electron).

potential (*electric*) The electrical condition at a point (for example, the e.m.f. of a cell) that can cause electric charges to move. It is a relative condition – Earth's potential is set at zero.

potential difference (p.d.) The difference in electrical potential between two points on a circuit or in an electric field. For two points on a circuit, it is the work done in moving a unit charge from one point to another or the energy transferred per coulomb of charge passing between the two points.

potential energy The energy possessed by a body because of its position. $E = mgh$, where E is the energy, m is the mass of a body at h meters above the surface of the Earth, and g is the acceleration of free fall.

potentiometer An instrument used to give an accurate measurement of e.m.f. or potential difference. It consists of a length of wire along which a contact can be moved. When it is in series with the source under test and a galvanometer, the length of wire where there is no deflection on the galvanometer gives a measure of the e.m.f.

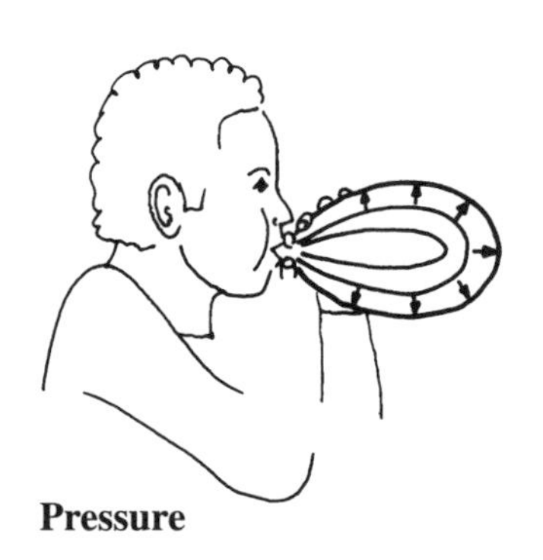

Pressure

power The rate of doing work or the rate at which energy is transferred.

precipitate An insoluble substance formed by a chemical reaction. Precipitation is the process by which a precipitate is formed.

pressure The force acting per unit area. In a liquid at rest, pressure increases with depth and acts equally in all directions.

primary cell *See* cell.

primary color There are three primary colors, red, green, and blue, in photographic reproduction. When light of these three colors is mixed, white light is formed.

principal axis The line that passes through the central point of the lens.

principal focus or **focal point** or **focus** The point on the principal axis through which rays of light parallel to the principal axis are refracted.

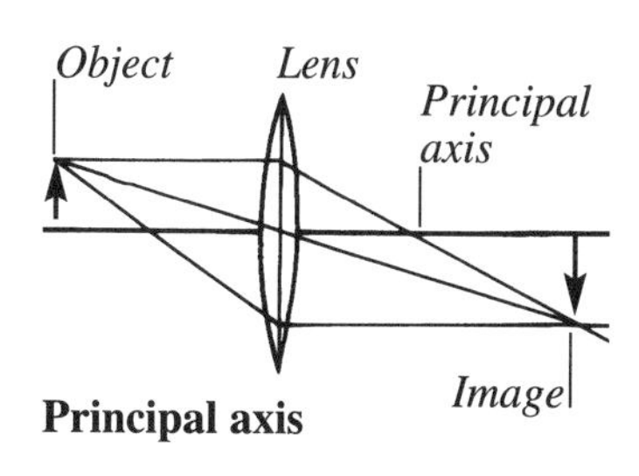

Principal axis

principle of superposition This is used to describe the interactions between waves. When waves meet, the effect of each wave adds up. When a wave at its point of maximum positive amplitude meets another of equal amplitude at its point of minimum amplitude, the result is zero – i.e., no light or sound or other phenomenon. If the maxima of two waves meet, the resulting wave has the height of both waves added together.

printed circuit An electronic or electrical circuit made by depositing a thin layer of copper on an insulating board so as to make the needed connections. Boards are often double-sided, and connections may be made through holes from one side to the other. Components, such as microchips, resistors, capacitors, inductors, transistors, diodes, and quartz crystals, are usually mounted on one side only but may be on both sides. These components are connected in circuit by soldering. Printed circuit boards (PCBs) are the standard electronic subunits in computers and may have various pronged plugs or sockets by means of which connections are made to power supplies and other parts of the machine. Alternatively, they may have multi-way edge connectors that can be pushed into long multi-way sockets. The main printed circuit board in a personal computer is called the mother board. Boards that can be plugged into this are called daughter boards.

prism A block of transparent material with triangular cross section. Triangular prisms are widely used in optical instruments and experiments. Prisms can be used to disperse light into a spectrum of colors. Ninety-degree prisms are used to give total internal reflection of light, which means that such a prism can bend light rays 180° in optical instruments, such as binoculars, to make the tubes shorter.

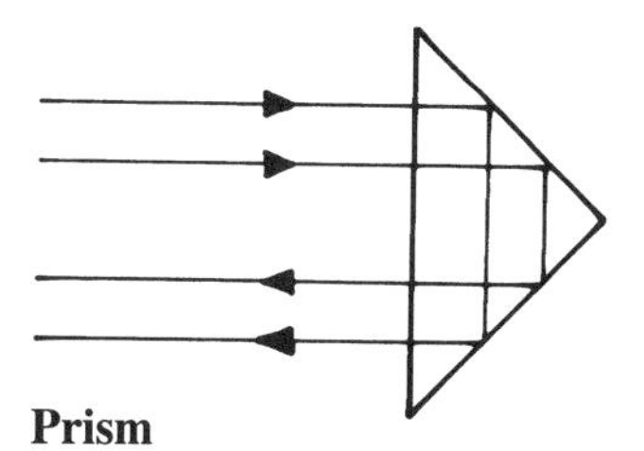
Prism

program A set of instructions or statements that can be represented in a form capable of being read by a computer and that cause the machine to perform specified actions. A program may be only a few bytes or many megabytes long.

programmable ROM (PROM) A read-only memory, the contents of which are entered after manufacture and may or may not, thereafter, be able to be changed.

programming All the activities, analytical, creative, and evaluative, involved in producing a computer program. The final stages of programming involve the coding and testing of the program and the detection and, if possible, elimination of faults and defects (debugging).

PROLOG A computer programming language based on formal logic and used to write artificial intelligence programs.

PROM *See* programmable ROM.

String
Pulley
Load

Pulley

proton One of the basic particles in the atom, found in the nucleus with the neutron. It is one of the most massive of the subatomic particles, similar in mass to the neutron. It has a positive charge. In a neutral atom the number of protons is equal to the number of electrons. Its mass is 1.673×10^{-27} kg.

proton number *See* atomic number.

pulley A simple machine for raising loads, consisting of a grooved wheel over which a rope or chain passes.

pulsar A collapsed neutron star that emits bursts of cosmic radio emissions in rapid pulses.

pulse In electronics and computing, a brief and quickly reversed change in voltage or current, usually such as to produce a waveform that is roughly rectangular.

pulse code modulation (PCM) Analog to digital conversion of an electrical signal in real time. The analog signal is sampled many times a second, and the amplitude of each sample is measured and represented as a number, which is then sent as a digital code.

pulse generator An electronic circuit with an electrical output, the waveform of which is a very short duration square or rectangular wave.

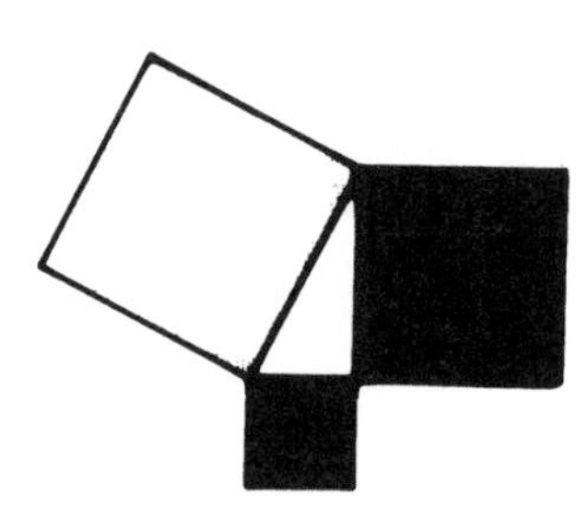

Pythagorean theorem

pyrometer An instrument for measuring high temperatures; the current through the filament in a bulb is adjusted until the bulb's color is the same as that of the hot body being studied. The measure of the current indicates the body's temperature.

Pythagorean theorem The area of the square on the hypotenuse of a right-angled triangle is equal to the sum of the areas of the squares on the other two sides of the triangle.

Q (Quality) The ratio of reactance to resistance in a resonant circuit. A high reactance and a low resistance means that such a circuit will resonate at a more precise frequency rather than over a wider frequency range. In other words, the greater the Q, the sharper the tuning of the circuit. A high Q inductance can be achieved by surrounding a low-resistance coil with a ferrite material that greatly increases the inductance without affecting the resistance.

quantum physics A fundamental theory in physics, proposed in 1900 by Max Planck, stating that the energy of a system can change only in small steps, or quanta, rather than having a continuous range of values. This theory was extended by Einstein in 1905 to electromagnetic radiation, which, it was suggested, travels in small energy packets, or photons. Quantum electrodynamics is a relativistic quantum theory of electromagnetic interactions that describes the interactions of electrons, muons, and photons, giving the underlying theory of electromagnetic phenomena. A system can have only certain possible quantum states or quantum numbers. The energy of a system is proportional to its frequency, and as this does not change continuously, the frequency must also change in whole number (quantum number) steps.

quark The basic fundamental particle forming the nuclei of atoms. Quarks combine to form hadrons, using the strong nuclear force.

quartz clock **1.** An accurate timepiece based on the piezoelectric properties of a quartz crystal incorporated into an electronic oscillator. Thus the natural resonant frequency of the crystal determines the frequency of the oscillator, which oscillates at this frequency with precision and stability determined only by the physical properties of the crystal. The output of the oscillator is then formed into a square wave train and electronically divided. When a suitably low frequency has been reached, the pulsed current is applied to a synchronous motor with a gear train and clock-face hands. Alternatively, pulses can be used to activate a digital display.
2. The quartz oscillator used in computers and other devices to synchronize the electronic activity.

radar Acronym for *ra*dio *d*etection *a*nd *r*anging, a method of locating objects, such as ships or aircraft, at a distance and determining their position. Radar works by projecting a pulsed beam of high-frequency radio waves of a wavelength short compared to the object to be detected. Such waves are reflected off the object and their returned echoes detected. Because the speed of propagation of radio waves is known, the distance can readily be computed from the time that elapses between emission and the arrival of the echo. The transmitted pulse and the returning pulse can be displayed as blips on a cathode-ray tube with a radially moving spot and a rotating radius whose position exactly mirrors that of the rotating aerial. The distance from the center of the display is a function of the distance of the object, and its bearing is a function of the position of the rotating transmitting/receiving antenna at the time.

radian The SI unit of angle. One radian is the angle subtended at the center of a circle by an arc equal in length to the radius of the circle; 2 π radians are equal to 360°.

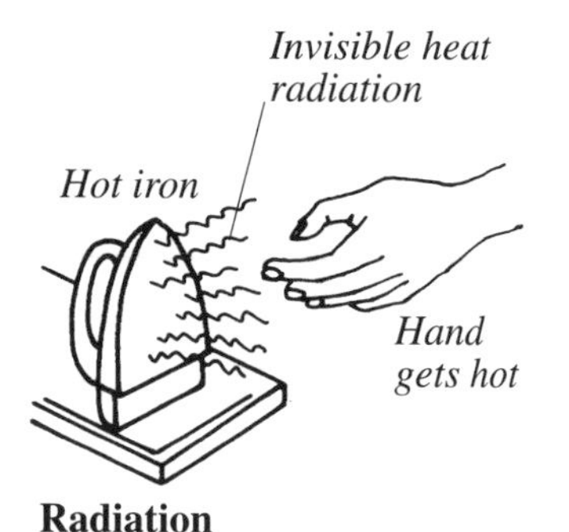

Radiation

radiance The objective physical property corresponding to the subjective sensation of the brightness of a surface.

radiation The transmission of energy, particularly electromagnetic radiation, by the movement of waves and particles in all directions. It does not require a medium in which to travel.

radiation pressure The very small pressure exerted on an irradiated object by the radiation. This pressure occurs because all radiation has momentum and energy.

radio The communication of information between remote points using electromagnetic radiation as the medium of transmission. Radio waves occupy the part of the electromagnetic spectrum between about 500,000 cycles per second (500 KHz) and 100 million cycles per second (100 MHz). Radio uses an alternating current carrier wave that is propagated through space in much the same way as light can be received direct or reflected from the ionosphere, about 37 miles (60 km) altitude, where there is a high concentration of free electrons. Imposed on the carrier wave is a modulation of various kinds, such as amplitude modulation (AM), frequency modulation (FM), pulse code modulation (PCM), and so on. When the modulated wave strikes an antenna, it induces a small alternating electric current in it, and this is greatly amplified by the radio receiver. The carrier wave is then removed, in a process known as demodulation, and the information retrieved. Most radio receivers use the superheterodyne principle.

radioactive decay constant (K) A measure of the probability of a nucleus decaying in the following second.

radioactive half-life The time in which half the atoms of a given quantity of a radioactive nuclide undergo at least one disintegration. An isotope's half-life is the natural logarithm of 2 divided by the isotope's radioactive decay constant.

radioactive tracers Labeling of nonradioactive material by adding small quantities of a radioactive preparation to study the movement of the material.

radioactivity The spontaneous disintegration of certain isotopes, accompanied by the emission of radiation (alpha rays, beta rays, gamma rays).

radiography The production of an image of a body structure by X rays.

radiometer An instrument used to detect and measure electromagnetic radiant energy (and acoustic energy). The bolometer and the thermopile are radiometers.

radius A straight line joining the center of a circle to a point on its circumference.

radius of gyration *See* moment of inertia.

RAM *See* random access memory.

random access memory (RAM) A high-speed semiconductor computer memory used as the main functional memory store into which programs and data in use are copied. RAMs are organized as two-dimensional arrays in which each cell can be independently addressed in a fixed access time and the contents read or overwritten. RAM may be static or dynamic. Static RAM maintains its information as long as the power supply is maintained; dynamic RAM must be automatically refreshed at regular intervals, otherwise the data would be lost. RAM is volatile, i.e., everything in it will be lost as soon as the power supply is switched off. As software develops, programs make increasing demand on RAM, and the recognized standard size has been increasing steadily. Currently, 64 million bytes (megabytes) of RAM are considered adequate, but for high-resolution images, greater RAM capacity is required.

rarefaction The opposite of compression for longitudinal waves such as sound.

Rayleigh criterion The criterion for the resolution of interference fringes, spectral lines, and images. The limit of resolution occurs when the maximum intensity of one fringe or line falls over the first minimum of an adjacent line or fringe.

Rayleigh's scattering law When electromagnetic radiation is reflected off atoms with unchanged energy and momentum, there is a change in phase but not in frequency.

reactance The impedance that a capacitor or an inductor offers to alternating current. For a pure inductor or capacitor, it is the ratio of the voltage across the device to the current through it at any moment in time. Reactance is measured in ohms.

reaction A process in which substances react to form new substances. Bonds are broken and re-formed in chemical reactions.

read only memory (ROM) A computer memory, the contents of which are permanently retained and are not lost when the machine power supplies are switched off. Most ROMs are smaller than most RAMs and contain operating programs required to get the machine started, as well as frequently used software. Although the contents of the ROM are readily accessible, they cannot normally be changed by the nontechnical user of the machine.

real time systems Computer systems that operate quickly enough for the result to be available for use in concurrent events in the real world without appreciable or unacceptable delay. The time from input to response in real time systems is often of the order of thousandths of a second. For many purposes, a longer delay is acceptable; for some, the response must be more rapid.

rectifier A device that allows electric current to flow freely in one direction but offers high resistance to flow in the reverse direction. Rectifiers, such as semiconductor junctions, are widely used to convert alternating current to unidirectional direct current. Rectified AC consists of a series of pulses, but these can easily be smoothed by capacitors into a direct current of uniform amplitude.

red giant A very large red-hot star that is formed when a white-hot star, such as the Sun, begins to run out of hydrogen. The star expands and cools down.

red shift This features in the spectra of astronomical objects, such as galaxies. The observed spectra of such objects have shifted toward longer wavelengths due to the Doppler effect because the astronomical objects have been moving away from Earth as a result of the expansion of the universe.

redundancy The provisions of physical components additional to the minimum needed for proper functioning. The purpose of redundancy is to ensure continued operation, even in the event of component failure. Redundancy is also used in software, especially in communications, to ensure that essential information is transmitted correctly.

reflecting telescope A telescope that brings light rays to a focus using a mirror. This overcomes the problems of chromatic aberration when lenses are used. The world's largest telescopes are all reflecting telescopes.

reflection The process by which radiation incident on a surface does not penetrate the surface but bounces back, obeying the laws of reflection.

reflection factor The ratio that light energy reflected per second from a surface makes with that falling on the surface.

refracting telescope A telescope that brings light to a focus using lenses. These tend to be the smaller telescopes used by amateurs.

refraction The change of direction of electromagnetic radiation, such as light, on passing from one medium to another. This occurs because the radiation travels at different velocities in different media.

refractive index The constant representing the amount of refraction occurring when electromagnetic radiation passes from one medium to another. It can be found using the velocities of light in the two media. The refractive index for light passing from medium 1 to medium 2 is equal to the velocity of the light in medium 1 divided by the velocity of light in medium 2. It is the ratio of the sine of the angle of incidence to the sine of the angle of refraction.

refractory Of a substance able to withstand high temperatures, making it suitable for use as lining material for furnaces.

refrigerator A device that uses the Joule-Thomson effect to maintain a storage area at low temperature.

relational database A database in which items of information are related to each other in some way so that information may be retrieved without undue duplication of the records. Information may be treated as subsets of other information and can be retrieved in accordance with logical formulas. Such a database, for instance, could find all instances that fulfill certain criteria and/or certain others but not stated others. Instances sought might also be equal to, less than, or more than stated values.

relative atomic mass The ratio of the mass of an average atom of an element

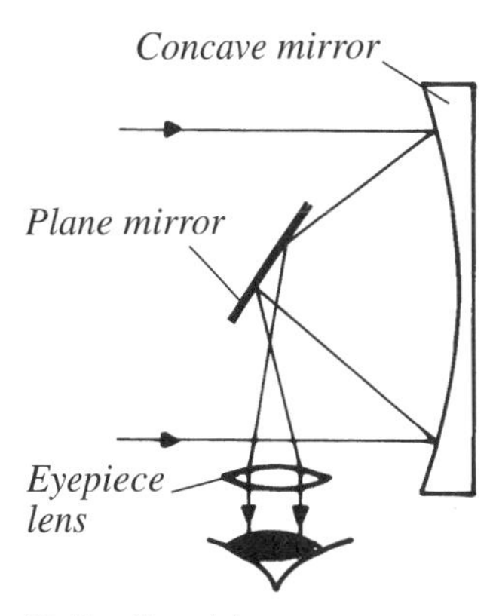

Reflecting telescope

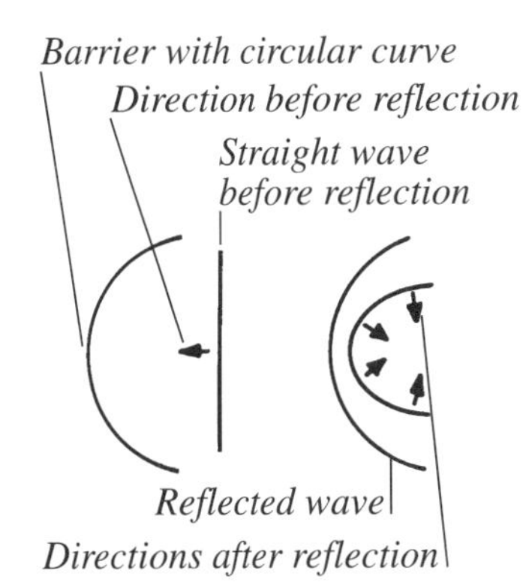

Reflection

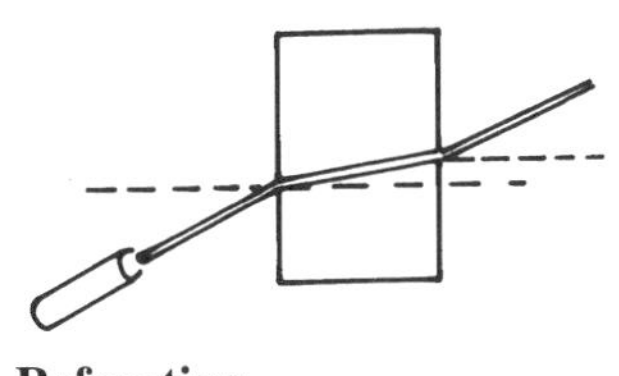

Refraction

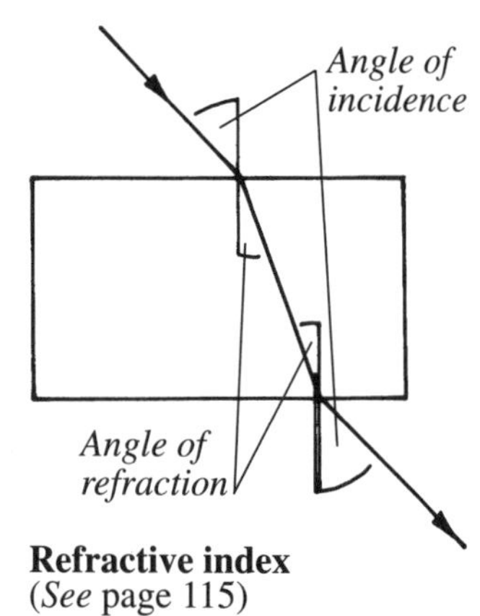

Refractive index
(*See* page 115)

to 1/12 of the mass of an atom of the carbon-12 isotope. (The mass of an atom of the carbon-12 isotope is taken as 12.)

relative molecular mass The ratio of the mass of a molecule of the element or compound to 1/12 of the mass of an atom of the carbon-12 isotope. The mass of an atom of the carbon-12 isotope is taken as 12.

relative permeability The ratio of magnetic flux density produced in a material to the value in free space produced by the same magnetic field strength.

relative permittivity (dielectric constant) The ratio of electric flux density produced in a material to the value in free space produced by the same electric field strength.

relativity, principle of The laws of mechanics, optics, and electromagnetics are the same in all uniformly moving frames of reference. The velocity of light in a vacuum is constant and does not depend on the motion of the observer. (If a process is observed from outside a moving frame of reference, it takes a longer time than that measured by an observer in the moving frame of reference.) The general theory of relativity predicts the curving of light near a massive object and is mainly concerned with gravity.

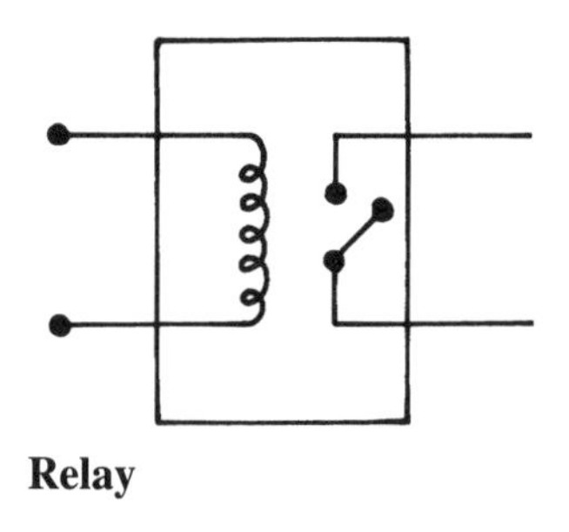
Relay

relay (*electrical*) An electromechanical switch that closes when an armature is attracted by a magnetic field caused by a current flowing through a coil.

resistance A measure of the extent to which a material can oppose the passage of an electric current. (It depends on the nature of the material, its dimensions, the temperature, and, sometimes, light.)

resistivity This gives a conductor's resistance in terms of its dimensions. A material's resistivity is the resistance of a one-meter length of the material, which is a square meter in cross section.

resolving power The potential an optical instrument has for producing distinguishable images. It depends on the wavelength of the light used and the aperture of the instrument.

resonance The state of a body capable of vibration or oscillation when it is subjected to a periodic disturbance at a frequency close to or equal to the frequency at which it naturally tends to vibrate or oscillate. At this frequency the body displays increased amplitude of oscillation or vibration, which may build up dangerously. The natural resonant frequency depends on material, dimensions, and shape. Resonance is also exhibited by electrical circuits. A coil (inductor) in parallel with a capacitor will resonate as the charge on the capacitor causes a current to flow in the inductor, and vice versa. This occurs at a frequency determined by the capacitance and the inductance and is the basis of radio tuning and many oscillators. Resonance occurs in molecules, atoms, and elementary particles.

respiration The process by which living organisms take in oxygen and give out carbon dioxide. It involves a series of chemical reactions that liberate energy for use in the body.

rest mass The property of a body that determines its inertia and its internal energy.

resultant The single vector representing the sum of a set of vectors.

reticule *See* graticule.

retina The light-sensitive area of the eye consisting of rods and cones.

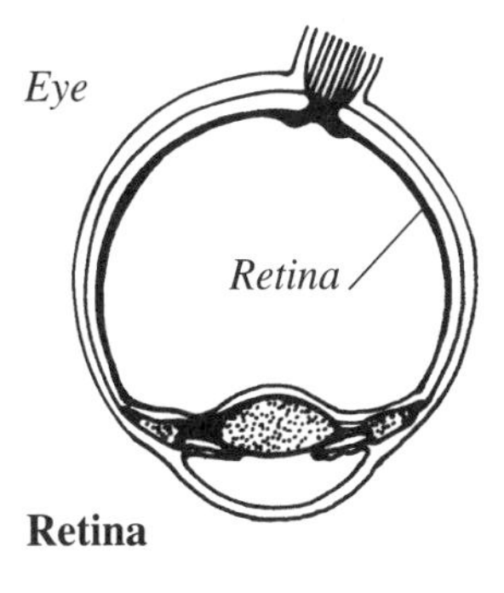

Retina

reverberation The effect of repeated reflection of sound by the walls of a large enclosed space after the sound that caused the original reflection has ceased. Moderate reverberation improves the acoustics of a room or building; excessive reverberation seriously interferes with discrimination of speech or music.

reversible reaction A chemical reaction that can proceed in either direction. It does not reach completion but achieves dynamic equilibrium.

Reynolds number A number that links the velocity, density, and viscosity of a fluid in a system of a particular length. The type of fluid flow is indicated by the number.

rheology The study of the flow properties of matter, especially viscous liquids.

rheostat A variable resistor.

Richter scale The scale used to indicate the magnitude of earthquakes. It measures the amplitude of a seismic wave on a scale in which an increase of one represents a 10-fold increase in the amplitude.

right-hand grip rule A technique to remember the direction of magnetic effects.

right-hand screw rule A technique to remember the direction of magnetic effects.

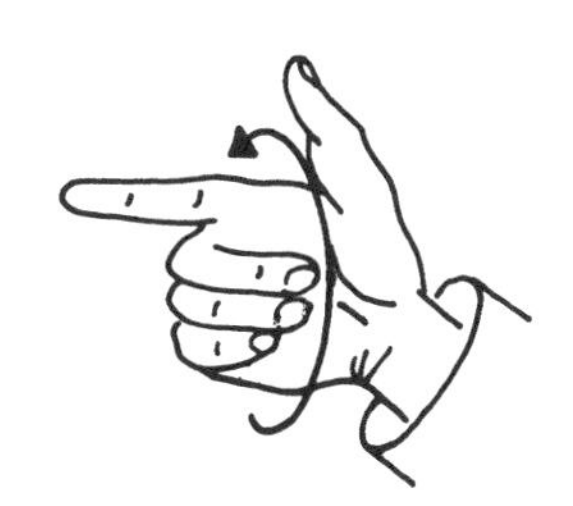
Right-hand screw rule

RISC Acronym for *r*educed *i*nstruction *s*et *c*omputer, a machine using a microprocessor with a minimal set of instructions so that it can operate very quickly. The necessary complexity is transferred to software.

rms value of alternating current The square root of the average of the squares of the values of the alternating current over a complete cycle. It is used to find the direct current equivalent of the alternating current.

rms value of alternating e.m.f. The square root of the average of the squares of the voltage taken over a complete cycle.

rocket A device for producing thrust by the rapid conversion of a liquid or solid fuel into hot and expanding gases. The combustion of fuel takes place with internally stored oxygen. The gases produced by combustion exert pressure on all parts of the interior of the combustion chamber, but because the rear end is open, the force exerted on the front of the chamber is unopposed, and the rocket moves forward. The escaping gases do not press against anything; rockets work best in a vacuum.

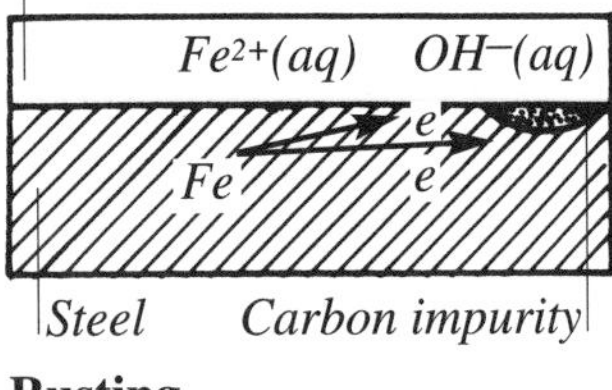

Rusting

ROM *See* read only memory.

rotational frequency The number of rotations per unit time.

routine A discrete part of a computer program that achieves a particular task. Other parts of a program may repeatedly call on routines.

rusting The way in which iron is attacked by air and water to form rust (hydrated iron oxide) on its surface.

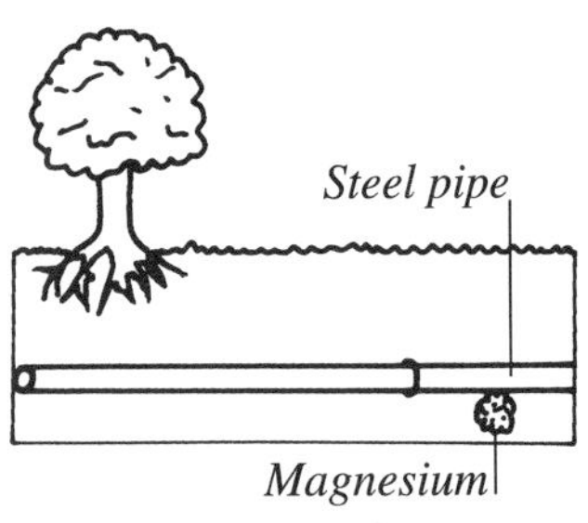

Sacrificial protection

sacrificial protection By attaching a more electropositive metal to the metal that requires protection, the protected metal is no longer corroded because the attached metal has become the anode in the corrosion process and is corroded in its place.

salt A compound formed from an acid when all or part of the hydrogen atoms are replaced by a metal or metal-like group. They are generally crystalline.

satellite A small body orbiting a larger one under the influence of gravity.

saturated A solution in which there is an equilibrium between the solution and its solute.

saturated compound An organic chemical compound that contains no double or triple bonds.

sawtooth waveform A periodic electrical wave, each cycle of which rises steadily from zero to a particular maximum over the course of almost the whole cycle, and then suddenly drops to zero at the end of the cycle. Also known as a ramp waveform. Two sawtooth voltage waveforms applied to the electron beam in a cathode-ray tube, one for horizontal spot movement, the other for vertical movement, are the basis of television and computer displays.

scalar A quantity that has magnitude but not direction.

scanning electron microscope An electron microscope that causes a very fine beam of electrons to scan the surface of the sample. Electrons are emitted at each sample point, depending on the properties of the sample at this point, and the electrons collected are manipulated to form an image.

scanning tunneling microscope An instrument for examining surfaces at a molecular and even atomic level. A very fine metal tip is positioned within a distance from a surface equal to the diameter of a few atoms. At such distances, electrons can jump (tunnel) across the gap. A movement of electrons is an electric current, and this varies with the distance from the surface. The changing current is amplified and used to adjust the tip's distance from the surface so as to keep the current constant. The up and down movement of the tip thus reflects the contours of the surface, and can do so to a resolution equal to individual atoms. If the tip is kept close to the surface, individual atoms can be moved to a new position.

scattering The deflection by any material (solid, liquid, gas) of radiation that strikes it. High-frequency radiation (at the blue end of the spectrum) is scattered more than lower frequencies.

Schmidt camera A camera using a spherical mirror and a correcting plate to compensate for the spherical aberration of the mirror instead of a lens. Such a system allows large apertures, freedom from chromatic and spherical aberration, and long focal length. The camera is used for astronomical and general purposes.

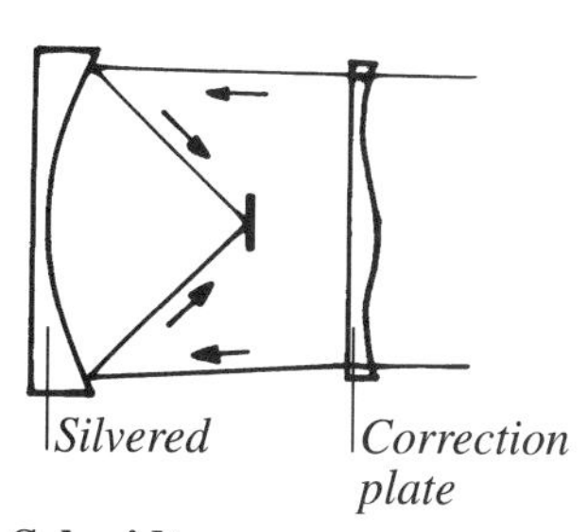

Schmidt camera

Schmitt trigger An electronic circuit whose output can be one or other of two voltages – high or low, corresponding to the 1 or 0 in binary electronic systems. If the input is above a certain level, the output will be high; if below that level, low. A sinusoidal voltage applied to a Schmitt trigger thus gives a sharp square wave output. All digital electronic systems work on square waves. The Schmitt trigger is an essential element for restoring sharpness when square waves have been degraded or corrupted by interference, or other causes.

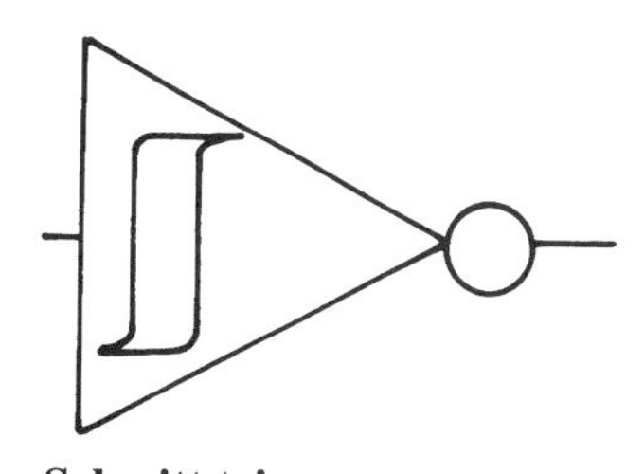

Schmitt trigger

Schottky diode A metal semiconductor, two-electrode circuit element with a low forward-bias voltage drop – about half that of other semiconductor diodes – and a higher switching speed. Integrated circuit transistors can be made incorporating Schottky diodes.

Schuler pendulum Anything that swings under the influence of gravity with a period of 84.4 minutes. This is the period of a pendulum equal in length to the radius of the Earth, and it has been shown that such a pendulum will remain vertical however the pivot may move. Gyroscopes can slow a natural pendulum of practical length so that its period is increased to 84.4. minutes, and these are used for navigation. Inertial navigation is based on the principle of the Schuler pendulum.

scintillation counter A detector of radiation in which the receipt of a quantum of radiation is signaled by a flash of light that is usually detected and amplified by a photomultiplier tube. The basis is a material, usually crystalline, that emits a flash of luminescence when struck by X rays, gamma rays, or high-speed particles. Scintillation counters also quantify the scintillation, so are able to measure the energy of the radiation.

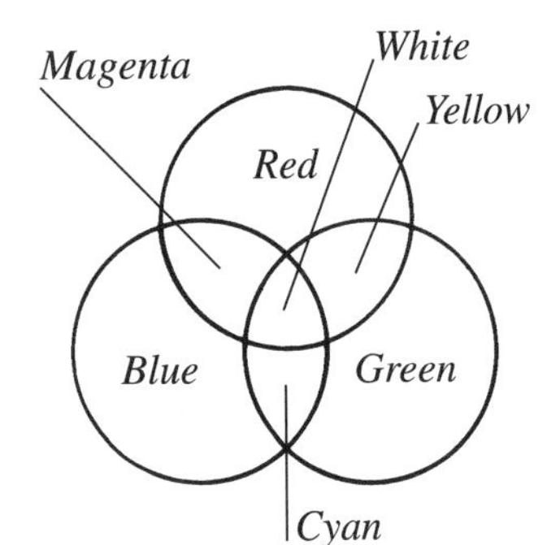

Secondary color

scratchpad memory A semiconductor memory of small capacity but very short access time, used to store intermediate results in the course of computation.

secondary cell *See* cell.

secondary color Magenta (red), cyan (blue), yellow. Each is formed by mixing the light of two different primary colors.

secondary emission The emission of electrons from the surface of a solid when it is bombarded by high-speed particles, especially electrons. Under certain circumstances, the number of electrons emitted may exceed the bombarding number, and this fact can be used to make photomultipliers and other instruments of great value. Secondary emission occurs most readily with insulators or semiconductors such as cesium antimonide, potassium chloride, beryllium oxide, and magnesium oxide.

sedimentation The settling out of particles in suspension in a liquid at the bottom of the liquid, because of gravity.

Seebeck effect The flow of electric current in a circuit consisting of two wires of different metals twisted together to form a ring when the two junctions are at different temperatures. The direction of current flow depends on which is the hotter junction. The effect was discovered in 1821 by Russian-born German physicist Thomas Johann Seebeck. *Compare* Peltier effect, in which a current caused to flow through such a circuit by a battery produces a raised temperature at one junction and a lowered temperature at the other.

selectivity The ability of a radio receiver to abstract a signal with a comparatively narrow frequency bandwidth from the spectrum of transmitted frequencies and to reject the others, especially those of similar frequencies. Selectivity is achieved by the use of several cascaded sharply tuned circuits that offer a high ratio of reactance to resistance (i.e., a high Q). The heterodyne principle facilitates the practical duplication of successive tuned circuits.

self-inductance If the current in a circuit changes, the magnetic flux linked to the circuit changes and induces an e.m.f. whose direction opposes the change causing it. *See* Lenz's law.

semiconductor A material whose resistance is between that of insulators and conductors. Semiconductors are covalent crystals. Current is carried through a semiconductor by free electrons, leaving positively charged holes behind. A pure semiconductor has equal numbers of holes and free electrons in its structure. The addition of impurities can alter the relative numbers of holes and free electrons *(see* acceptor), creating n-type semiconductors that have an excess of electrons and p-type semiconductors that have an excess of holes.

semiconductor diode A two-connection circuit element consisting either of a junction between a conductor and a semiconductor or a junction between p-type and n-type semiconductor materials. Such a diode allows current to flow in one direction only. Semiconductor diodes with p-n junctions can be arranged to function as light sources (light-emitting diodes or LEDs), as switches, as voltage regulators *(see* Zener diode), or as variable capacitors (varicaps).

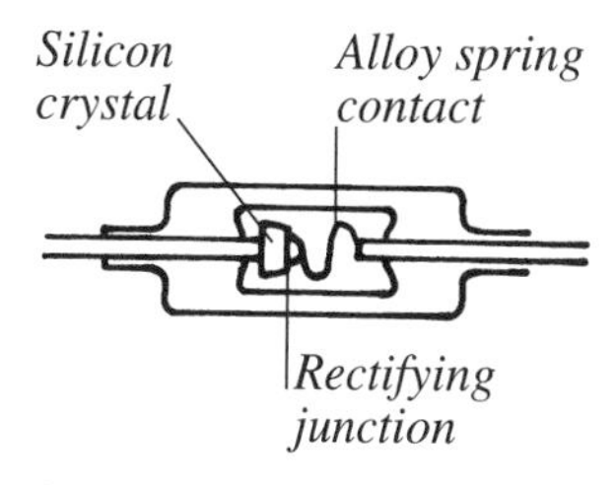

Semiconductor diode

semiconductor laser A device in which laser action occurs through the stimulated recombination of free electrons and holes in the valency band in a semiconductor crystal such as gallium arsenide. In so doing, energy is released and radiated as a quantum of light. A current passed through a p-n junction in such a material results in large numbers of electrons and holes being brought together. The flat ends of the crystals act as mirrors to promote lasing. Almost every personal computer and CD player now contains a semiconductor laser.

semiconductor rectifier A semiconductor device for converting an alternating current into a direct current by allowing current to flow in one direction only. Silicon p-n junctions are used, and these can pass currents of less than 1 amp to as much as 5,000 amps. Diodes can be connected in parallel to carry greater current or in series to resist higher voltages. A third electrode can be introduced to exert a controlling effect on the passing current. These are called silicon-controlled rectifiers (SCRs) or thyristors and are extensively used as power-control components, such as light dimmers.

serial and parallel transfer Alternative ways of sending binary information. In serial transmission, the bits are sent one after the other. This can be done using a single line and a return line. In parallel transmission, an eight-bit byte (or longer word) is sent along eight (or more) parallel channels, the bits moving together. Parallel connection is commonly used to send data to a printer, but, for economic and other reasons, is rarely used for long-distance transmission.

serial interface A connection at a computer at which digital information passes out or in one bit at a time. Serial interfaces can convert parallel data to serial, and vice versa. A common standard for serial interfaces is the RS-232 standard.

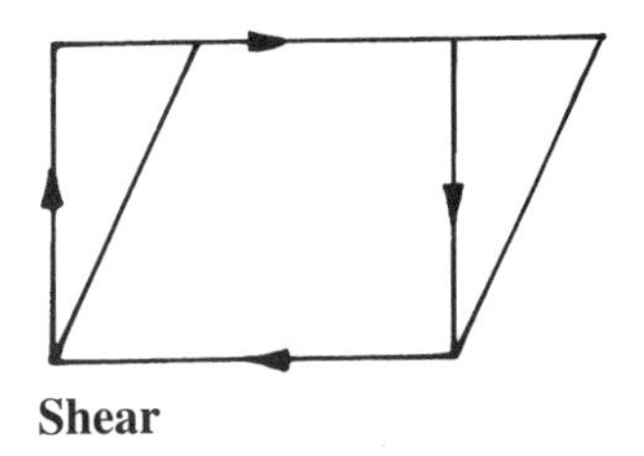

Shear

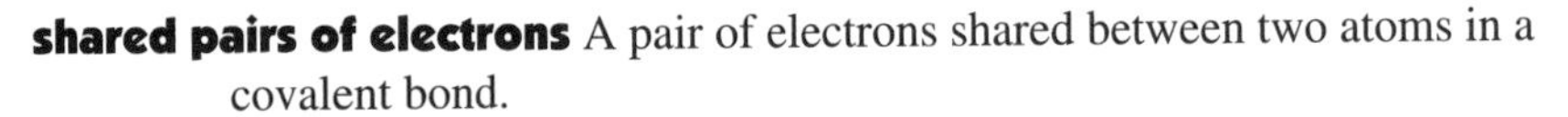

shared pairs of electrons A pair of electrons shared between two atoms in a covalent bond.

shear When a material's structure of parallel planes becomes displaced in a direction parallel to themselves, there is a tendency for adjacent planes to slide over each other, causing deformation of the material.

shear modulus Ratio of shear stress to shear strain.

shear strain The distortion of an object by a force acting across the section of a beam loaded transversely (or across shaft sections subject to torque). The value of the shear strain is the ratio of the distortion to the original dimension.

shear stress Force per unit area acting across the section of a beam loaded transversely (also across shaft sections subject to torque) and tending to change its dimensions (i.e., to cause a strain). The size of the stress varies across the section and is dependent on the geometric shape of the body.

shift register A small computer memory consisting of a ladder-like array of places, each of which can accommodate a whole eight-bit or larger word. With the arrival of each clock pulse, the whole word is shifted along one place, the end word being ejected into a bus and a place being left, at the other end, for a new word to enter. Shift registers are widely used in computing and may be implemented either as actual hardware or as software.

short circuit An electrical connection of negligible resistance that bypasses a part of an electrical circuit. Because the resistance is so low, a maximal current will flow though a short circuit, and this may have a dangerous heating effect.

shunt Anything that bypasses a normal route. Sensitive electric-current meters in series with a conductor carrying a heavy current are always bypassed by a very low-resistance parallel conductor, so that only a small but fixed proportion of the current passes through the meter.

SI Abbreviation for *S*ystème *I*nternational (d'Unités), a system of coherent metric units (SI units) proposed for international recognition in 1960 and now the standard system of units used in science.

signal reception The reception of a radio signal by a receiver. The wave is transmitted as a radio frequency carrier whose amplitude has been modulated by an audio frequency. The signal is picked up by an antenna, causing free electrons in the antenna to oscillate, producing an alternating current of the same frequency as the radio signal (induced by the electric field component). The radio's ferrite rod antenna responds to the magnetic component of the radio wave and induces a current in the coil. It is tuned to allow the correct frequency of signal to be demodulated to remove the radio frequency carrier.

sine wave The projection on a plane surface of the graph of a point moving around a circle at uniform speed, and the graph of the function $y = \sin x$ or $y = \cos x$. It is also the projection onto a uniformly moving plane of a point on a pendulum swinging above it. Sine waves of different frequencies and amplitudes are the waves into which any other periodic waveform, however complex, can be analyzed by the process of Fourier analysis. A sine wave has no harmonics and an acoustic sine wave sounds pure and rather dull, like a low note on a flute played softly. Domestic alternating current is sinusoidal.

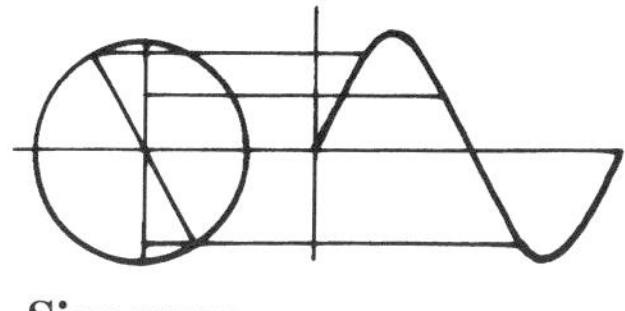

Sine wave

singularity A hypothetical point in space at which the general theory of relativity does not apply because matter is infinitely compressed to a single infinitesimal point. Current theory indicates that this was the situation immediately before big bang and is the situation in a black hole.

slag Waste material that collects on the surface of a molten metal during the process of either extraction or refining. It is composed of oxides, phosphates, silicates, and sulfides.

slit separation The measure of the gap between slits in a system used to study diffraction (at a double slit).

slope The inclination of a line compared with a fixed line (usually the horizontal). On a graph, it is the tangent of the angle made between the line and the horizon.

Snell's law The ratio of the sine of the angle of incidence to the sine of the angle of refraction is a constant for a given pair of media (the refractive index).

software The programs of instructions or the collections of data used by computers. Software, whatever its usable content, consists of sequences of binary numbers representing instructions and data and is thus intangible. These numbers are usually preserved in the form of tiny magnetic points on a magnetizable medium or, more permanently, as microscopic pits on the surface of a compact disk. These disks or other media are not software; it is the information on them that constitutes the software. By association, however, the disks are sometimes referred to as software.

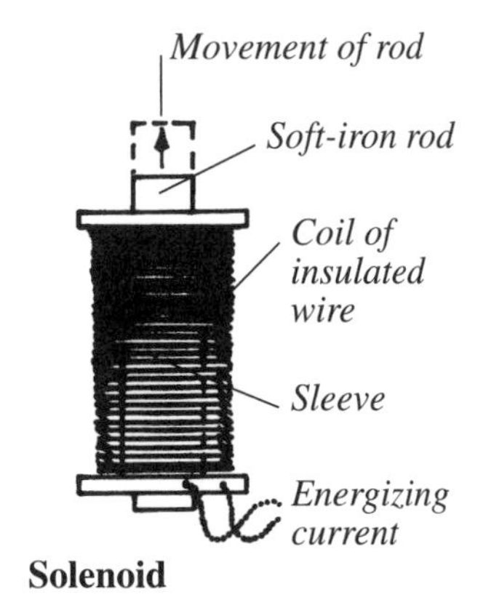

Solenoid

solar system The astronomical system comprising the Sun and all its attendant bodies, major or minor (planets, asteroids, comets, meteors). It can also refer to another star and its attendant bodies.

solar wind A continuous stream of particles, mostly protons and electrons, emitted from the surface of the Sun at supersonic speed.

solder An alloy used to join metals; it contains different metals depending on the requirements.

solenoid A coil of insulated wire, cylindrical in shape, with its length longer than its diameter. A magnetic field similar to that produced by a bar magnet is produced when an electric current is passed through the coil of wire.

solid state An electronic device that uses semiconductors.

solute A substance that dissolves in a solvent to form a solution.

solution A uniform mixture of one or more solutes in a solvent. It usually refers to solids dissolved in liquids but can also refer to gases in liquids, gases in solids, etc.

solvent A substance, usually a liquid, in which a solute dissolves to form a solution.

sonar Acronym for *so*und *na*vigation and *r*anging, a system used to detect and locate objects (similar to radar), using reflected sound waves. It is used underwater, where radar cannot work.

sonometer An apparatus that is used to study the vibrations of a wire or string.

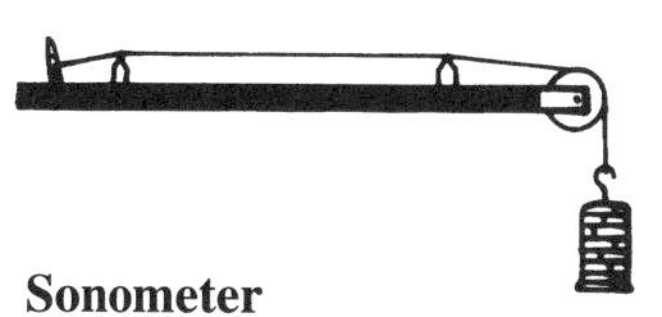

Sonometer

sound absorption The process in which sound energy is diminished by its conversion into heat in the course of its interaction with matter. Soft, nonrigid and easily deformed material is a more efficient absorber of sound than hard, reflective material.

sound energy Energy that is transmitted through a solid, liquid, or gas by longitudinal waves as a series of alternate compressions and rarefactions of the medium.

sound intensity (loudness) The flux of sound power through a unit area normal to the direction of propagation.

sound waves *See* sound energy.

spark chamber Radiation detector consisting of a chamber containing a stack of parallel metal plates with a strong electric field between them. Ionizing particles entering the chamber are made visible by the sparks formed after ionization.

specific charge The charge-to-mass ratio of an elementary particle. Its value assists identification of the particle being investigated.

specific charge of electron The charge-to-mass ratio of the electron: 1.8×10^{11} C/kg.

specific gravity The ratio of the density of a material to the density of some standard material, often water at a temperature of 15°C (60°F). Density is mass per unit volume.

specific heat The ratio of the amount of heat required to raise the temperature of a given weight of a material by one degree to the heat required to raise the same weight of an accepted standard material, usually water, by the same temperature and starting at the same temperature.

specific latent heat (1) (*fusion*) The heat required to convert unit mass of the substance from solid to liquid.

(2) (*vaporization*) The heat required to convert unit mass of the substance from liquid to gas.

specific volume The volume of unit mass. It is also the reciprocal of density.

speckle The patterns of random intensity light sparkles produced when a laser beam is reflected from a rough surface. This pattern is caused by many coherent wavelets that interfere with each other to form a granular pattern on the surface of most objects viewed by coherent light. Speckle has been applied scientifically in measurement (metrology) and in speckle interferometry in astronomy.

spectral lines Distinct lines, each representing a certain wavelength of radiation.

spectrograph An instrument that makes a photographic record, a spectrogram, of a spectrum (used in analysis of a substance).

spectroscopy The study of spectra.

spectrum The arrangement of electromagnetic radiation into its constituent wavelengths. A well-known spectrum is that of visible light, which consists of the colors red, orange, yellow, green, blue, indigo, violet.

speech recognition Computer analysis of an acoustic speech signal, received by a microphone, so that linguistic and sometimes semantic information can be extracted from it. The computer may have to be "taught" to recognize the user's pronunciation. More complex systems may be able to distinguish from the context the different meanings of phonetically identical words. Such systems necessarily involve artificial intelligence.

speed The distance moved in unit time by an object or particle. It is a scalar quantity.

speed of electromagnetic waves All electromagnetic waves travel at 3×10^8 meters per second in a vacuum. Nothing can travel faster than this.

speed of rotation The number of rotations of a rotating body about its axis of rotation in unit time.

speed of sound The speed of sound varies depending on the medium of transmission and the temperature; the more dense the medium, the faster the speed of sound within it.

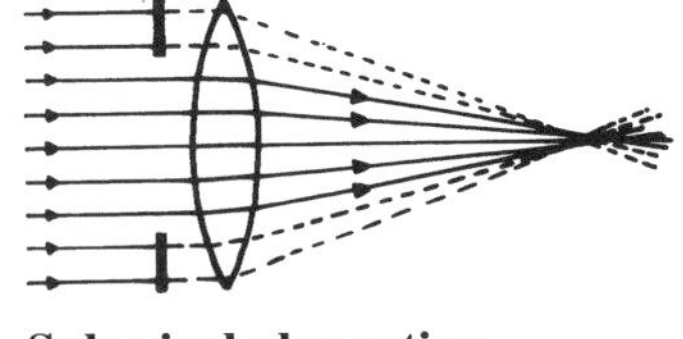

Spherical aberration

spherical aberration The distortion in the image formed by a lens or mirror resulting from the fact that rays striking a lens or a mirror at different distances from the center do not necessarily come to a focus at the same point. Spherical aberration can be reduced by the use of an iris diaphragm to "stop down" the system. *See* also astigmatism.

spin In quantum mechanics, the inherent angular momentum (circular motion around an axis of rotation) of a particle when it is at rest, distinguished from the orbital angular momentum of the particle. A particle in a particular energy level has a definite spin, and this is one of its intrinsic characteristics.

square wave An electrical waveform that alternates between a high and a low voltage, changing very suddenly from one to the other. The high voltage is taken to correspond to a binary 1 and the low voltage to a 0. In computers a square wave is produced by a quartz-crystal-controlled oscillator and is maintained by a wave-shaping circuit, such as a Schmitt trigger. The train of square waves forms the synchronizing clock signal by which the whole machine is kept in step. Transistors are operated in a switching, on-off mode that essentially produces square waves; the switching itself is effected by square wave voltages applied to the transistors.

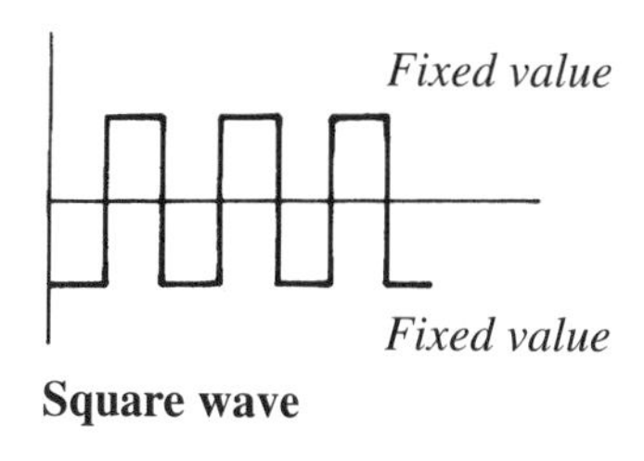

Square wave

Stark effect The splitting of emission spectrum lines when the source is subjected to a strong electric field.

states of matter The four states are solid, liquid, gas, and plasma.

statics The branch of mechanics concerned with bodies acted on by balanced forces so that they remain at rest or in uniform motion.

stationary wave (standing wave) The interference pattern formed by two waves of the same type that have the same frequency and amplitude and are traveling at the same time in opposite directions. A stationary wave form is produced, showing antinodes (points of maximum displacement) and nodes (points of zero displacement).

steam engine An external combustion engine. A boiler produces steam, which is introduced into cylinders in which pistons move.

Stefan-Boltzmann law The total energy radiated from a black body per unit area per second is proportional to the fourth power of the absolute temperature. $E = \sigma T^4$, where σ is the Stefan-Boltzmann constant.

stellar parallax *See* parallax.

steradian The SI unit of measurement of solid angle. One steradian is the angle subtended at the center of a sphere by an area of its surface numerically equal to the square of its radius.

stochastic process Any process controlled by the rules of probability. The hardness of a table top is determined by the probability that atomic movement will continue to result in a net homogeneity and will not result in a local deficit.

STP Conditions of standard temperature and pressure.

strain A body's temporary or permanent distortion when a stress is applied to it. It is given as the ratio of the body's change in length, area, or volume to its original dimensions.

strain gage Any device that measures strain. Strain gauges are attached to objects likely to experience strain in such a way that the gauge itself suffers strain. By various means, such as change in electrical resistance, capacitance, or inductance, this deformation is converted into an electrical change that can conveniently be measured.

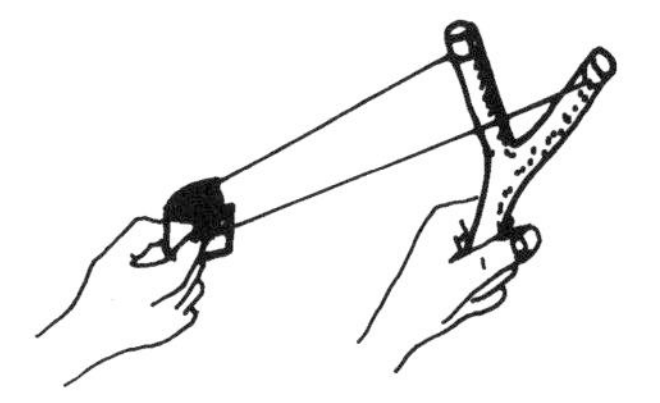

Strain energy

stratosphere Part of Earth's atmosphere, between the troposphere and mesosphere.

stress A force producing distortion in a body. It is measured in force per unit area.

string In computing, any sequence of characters. Strings may be searched for, compared, matched, replaced, joined (concatenated), stored, retrieved, or manipulated in other ways.

stroboscope An instrument used to study vibrating or rotating objects. It consists of a flashing lamp whose period can be adjusted to synchronize with the frequency of the object being studied so that the object appears stationary.

sublimate The solid substance that forms during sublimation – the reversible process by which a substance in a solid state changes directly to a gas. This process can be used to purify a substance.

substitution reactions A typical type of reaction for saturated organic compounds. One or more atoms, or groups of atoms, are replaced by other atoms or groups of atoms.

subtend The part of an arc of a circle (or a straight line) intercepted by and between the arms of an angle is said to subtend the angle.

summing amplifier An operational amplifier used to add a number of input voltages. It is used in sound mixers and amplifiers.

superconducting devices Devices able to carry out functions, by virtue of being superconducting at very low temperatures, which would not be possible or economical with normal conductivity. They include high-speed semiconductor switches with a switching time of less than 1 nanosecond; microwave radiation detectors; magnetic flux change detectors; superconducting quantum interference devices (SQUIDs); large and powerful magnets for magnetic resonance imaging machines (MRI scanners), high-energy particle accelerators, controlled fusion reactors, and other purposes; highly efficient electric generators and superconducting electric motors.

superconductivity The absence of electrical resistance in electrical conductors cooled to temperatures close to absolute zero.

supercooled Cooled below the normal boiling or freezing point without becoming liquid or solid, respectively (i.e., without change of phase).

supercooled vapor A substance that exists as a vapor at a temperature below that at which it should have become liquid.

superfluidity Describes the behavior of helium below 2.19K. Its viscosity effectively vanishes and the atoms are in their lowest quantum state. As the temperature decreases, the proportion of superfluid helium increases.

superheated water Water at a temperature above that of water boiling at one atmosphere.

supernova explosion An exploding star rapidly flaring to about 100 million times its original brightness. The star is then either totally destroyed or becomes a neutron star.

surface tension Within a liquid, molecules attract each other equally in all directions. At the surface, however, there is no force attracting them outwards, so the molecules are pulled toward the interior of the liquid. For this reason, liquid surfaces tend to become as small as possible.

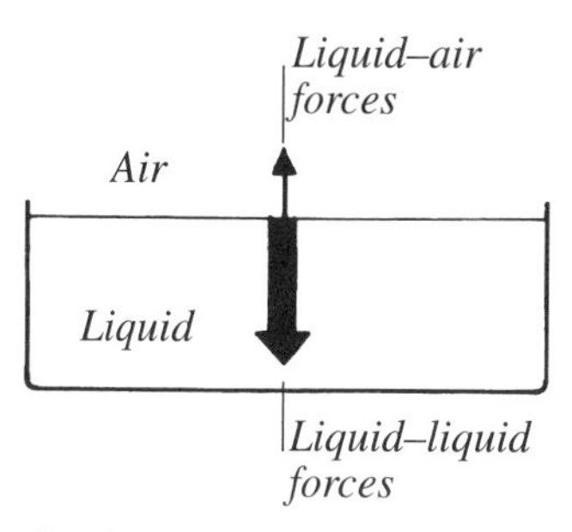

Surface tension

suspension A type of dispersion in which small solid particles are dispersed in a liquid or gas.

switching circuits Electronic circuits capable of rapidly moving from the "on" to the "off" condition of a small electric current. All digital electronics, including all digital computing, is based on switching of currents whose presence denotes a 1 and whose absence denotes a 0. It is thus important that switching should be as rapid as possible. Today, specialized switches can change state in less than

1 nanosecond (less than one thousand millionth of a second). *See also* Schmitt trigger, square wave.

synchronization Maintaining two regularly repetitively moving (periodic) systems in step with each other, either on a one step-to-one step basis or on the basis of some other ratio of steps. Synchronization is an essential feature of many electronic devices. Sync signals are transmitted with the video television signal, for instance, so that the line and frame oscillators that move, respectively, the electron beam horizontally and vertically are kept running at just the right speed to ensure that the spots on the screen correspond exactly to those on the television camera tube.

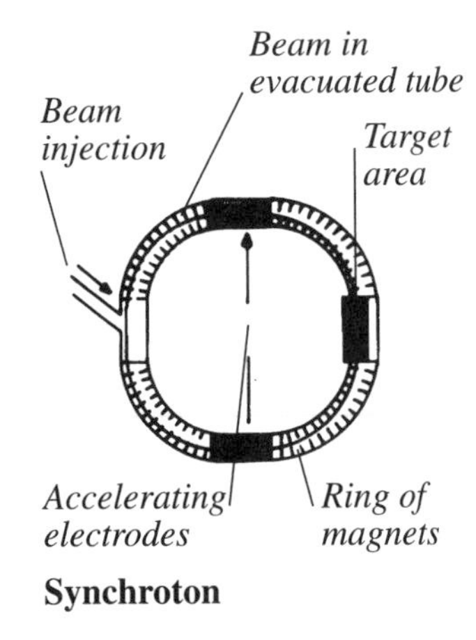

Synchroton

synchronous motor An electric motor that runs at a fixed speed by rotating at a number of times per second that bears a whole-number ratio to the frequency of the alternating current electricity supply. Prior to the introduction of quartz oscillator-controlled digital clocks the most accurate domestic clocks were those operating on this principle. Their accuracy was the same as that of the electricity mains frequency, which is usually kept within narrow limits.

synchrotron A type of particle accelerator in which electrical and magnetic fields combine to accelerate charged particles in a circular path.

synthesis The formation of chemical compounds by constructing them directly from their elements or from other simple compounds.

tachometer An instrument that measures rotational speed, such as that of a wheel or shaft, in revolutions per minute or per some other interval of time.

tachyon A theoretical particle that can travel faster then light by virtue of being massless.

tangent A line or plane that touches a curve or surface.

tarnishing The discoloring of an exposed metal or mineral surface.

tauon A negatively charged subatomic particle, a high-mass lepton.

telemetry The obtaining, over a period of time, of data at a remote or inaccessible site and its transmission to a convenient location, usually by radio. Telemetry has many applications. It may, for instance, be used to monitor internal conditions and changes from within the human or animal body, determine conditions on the surface of a distant planet or within a space satellite, or obtain continuous information about changes in various atmospheric parameters from a geographically remote site or a research balloon.

telescope An optical instrument enabling the magnification of distant objects.

television Transmission, reception, and reproduction of moving visual images by electronic means.

temperature A measure of the degree of hotness of a system on a particular scale – Kelvin scale, Celsius scale, etc.

tensile strength The stress required to break a material that is under tension.

tensile stress The force required to produce distortion in a body.

tension A body in equilibrium is under tension if a force is tending to stretch it.

terminal velocity The constant velocity acquired by a body falling freely through a fluid when acted on by a constant force, such as a body falling through air at a point where the force of gravity is equal to the frictional resistance of the air.

test tube A narrow glass tube, closed at one end, used in laboratories for experiments.

thermal conduction The transfer of heat energy through a substance from a region of high temperature to low. Energy is transferred by vibrations of adjacent molecules. The substance itself does not move.

thermal conductivity A measure of the ease with which a material conducts heat energy.

thermal equilibrium A position where there is no heat loss or gain for bodies within a system.

thermal neutron A neutron that is in thermal equilibrium with the moderator in a nuclear reactor. It moves with a sufficiently slow speed to undergo fission reactions.

thermal radiation The emission of infrared radiation.

thermal resistance Resistance to heat flow. The unit is the thermal ohm.

thermal sink The area to which heat is lost in the cycle of a heat engine. Also called a heat sink.

thermionic emission The ejection of electrons into an evacuated space by a heated electrical conductor. Prior to the development of the

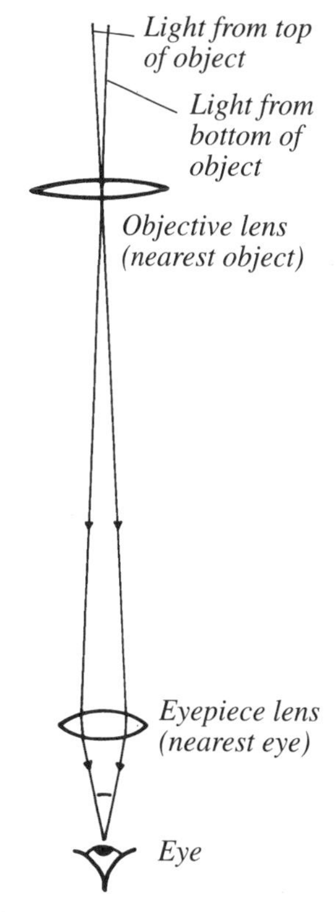

Telescope

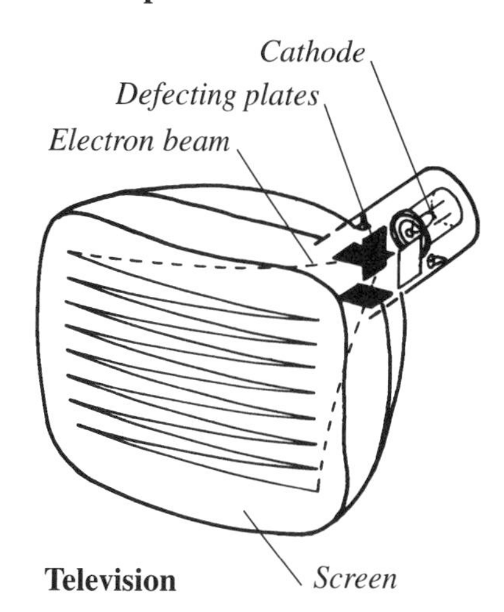

Television

transistor, thermionic emission was the basis of electronics. Thermionic valves are now obsolete, but the principle continues to be used in other devices, especially cathode-ray tubes.

thermionic valve (largely superseded by transistors) A glass or metal tube containing gas (or a vacuum) and two or more electrodes. The cathode is heated and the anode maintained at a positive potential. Other electrodes may be present to control the flow of electrons.

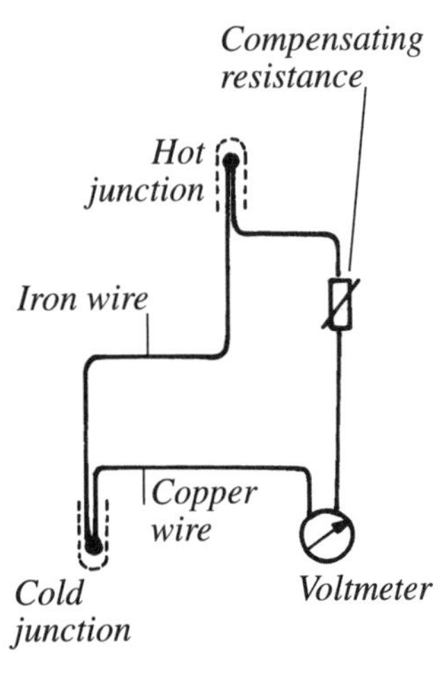

Thermocouple

thermistor (thermal resistor) A semiconductor whose resistance is very sensitive to temperature. It is used for temperature measurements and temperature warning devices.

thermocouple An instrument used for measuring temperature. It consists of a pair of wires or semiconductors joined at each end. One end is at the temperature to be measured while the other is at a fixed reference temperature. If there is a difference in temperature, a potential difference is produced, which results in a current that is proportional to the temperature difference between the two junctions.

thermodynamics The branch of science that deals with the relationship between heat and other forms of energy.

thermoelectricity The interchange of heat and electric energy.

thermometer An instrument used to measure temperature.

thermonuclear energy The energy produced by a nuclear fusion process. The reaction requires a high initial temperature, and the rate of reaction increases rapidly with increased temperature.

thermopile (pile) An instrument used to measure radiant heat. It consists of a number of thermocouples connected in series, with the alternate junctions exposed to receive radiant heat. The e.m.f.s produced by each junction are added together. This allows small temperature changes to be measured.

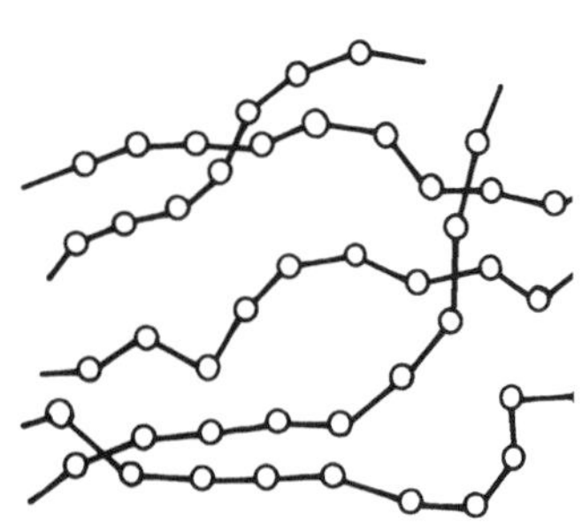
Thermosetting polymer

thermoplastic A substance (particularly a synthetic plastic) that becomes flexible when heated and hardens on cooling, with no change in its properties. Thermoplastic polymers have a molecular chain structure.

thermosetting polymer A polymer that has a structure of interlinked chains. Thermosetting polymers cannot be softened by heat but are decomposed by it.

thermosphere The region of the Earth's atmosphere above the mesosphere where temperature rises with height.

threshold frequency The minimum energy a photon must have to be able to release an electron from a surface.

thrust The propulsive force produced by a jet or rocket motor.

thyristor (silicon-controlled rectifier) A semiconductor power switch that has three electrodes – anode, cathode, and gate. A voltage is applied between the anode and cathode, and current flows when the anode is positive compared to the cathode – the gate is triggered. The current stops when the anode voltage drops to the cathode voltage.

tidal energy Energy represented by the large masses of water that move under the influence of gravity, following the pull of the Sun and Moon and forming tides. Electricity can be generated using the rise and fall of the large volumes of water involved in tidal movement.

titration The addition of a solution of known concentration from a burette to a flask containing a known volume of a sample of unknown concentration until the reaction between the two solutions is complete (this point is given by an indicator). The knowledge of the volume of liquid of known concentration added from the burette and the volume of liquid in the flask allows the concentration of the liquid in the flask to be calculated.

tonne Metric unit of mass: 1,000 kg.

torque The turning moment exerted by a force acting on an object at a distance from the axis of rotation and at a tangent to it.

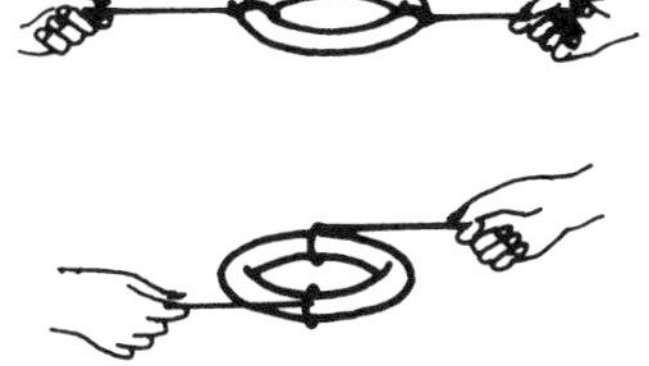

Torque

torsion The state of strain in a material caused by twisting.

torsion balance An instrument used to measure small forces, such as gravitation, magnetism, or electric charge, using the torsion produced in a thin wire.

transducer Any device that converts energy in one form, such as sound or pressure variations, light, heat, etc., into a corresponding electrical signal, or any device that can convert an electrical signal into corresponding changes in another form of energy. Examples of transducers are microphones, loudspeakers, photoelectric cells, piezoelectric pickups, ultrasound generators, and so on.

transformer A device that changes an alternating current of one voltage to an alternating current of another voltage by electromagnetic induction (mutual induction). It consists of two coils of wire wound onto a laminated iron core. The alternating current is applied to one coil, the primary coil, and current is induced in the secondary. If there are more turns of wire in the primary coil than in the secondary, the voltage induced will be lower than the voltage applied, and the transformer will be a step-down transformer. Similarly, a step-up transformer has more turns in the secondary coil than the primary coil and the resulting voltage in the secondary coil is higher.

transistor A semiconductor device that amplifies electric currents flowing through it. They usually consist of three layers of n-type and p-type semiconductors.

transition metals Metallic elements that have an incomplete inner electron structure. They are found in the central section of the periodic table.

transmission (electrical power) The way in which electric power is transferred to the place in which it is used, in the form in which it is required.

transmission factor (transmittance) The ratio of energy transmitted by a body to the energy incident upon it.

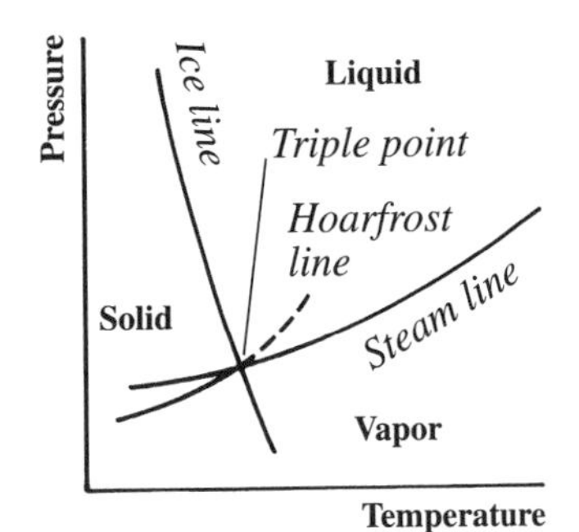

Triple point

transverse waves Wave motions where the vibration or displacement is perpendicular to the direction of propagation – for example, electromagnetic waves and water waves.

triple point The conditions of temperature and pressure at which the three phases of a substance – solid, liquid, and gas – are in equilibrium.

triple-point cell Apparatus for determining the triple point of water.

tripod An iron stand with three legs used in science laboratories to support apparatus.

troposphere The lowest of the Earth's atmospheric layers.

truth table The table that shows all the possible inputs and outputs for a digital circuit in which logical operations take place.

turbine A machine that extracts energy from a moving fluid – steam, air, water, gas. It consists of a shaft, fitted with blades, that is made to rotate by the force of the moving fluid. The rotation of the turbine shaft can be used to generate electricity.

turbulence (atmospheric) Fluid flow in which there is an irregular eddying motion leading to high drag.

tuyères Small jet holes at the base of a blast furnace through which preheated air is blown.

Tyndall effect The scattering of light by very small particles of matter – the scattered light is mainly blue. This is seen when sunlight passes through a dusty atmosphere.

ultrasound Sound waves with a higher frequency than that detectable by the human ear (above 20 kHz). It is used by bats for navigation, industrially, to check for faults, and medically, as a diagnostic tool.

ultraviolet light (UVL) Electromagnetic radiation of shorter wavelengths than visible light, but of longer wavelength than X rays. UVL is invisible to the human eye and is sometimes called black light. The spectrum of UVL is arbitrarily divided into three zones. That nearest to visible light (UVA) covers wavelengths from 380 to 320 nanometers (billionth of a meter); UVB ranges from 320 to 290; and UVC from 290 to one-tenth of a nanometer. UVC is especially penetrating and harmful to human tissue, but is strongly absorbed by the ozone layer in the Earth's stratosphere. Most of the UVB content is also filtered out by this layer.

uncertainty principle The theory proposing the limit to the accuracy of simultaneous measurement of energy and time, or position and momentum.

unified force At the very beginning of the universe, it is suggested that the four fundamental forces of nature formed a single unified force, supergravity.

universal gas law equation *See* gas equation, molar gas constant.

universal gravitation Any two bodies of matter attract each other with a force that is proportional to the product of their masses and inversely proportional to the square of the distance between them.

universe All space and the matter within it.

UNIX An important computer operating system widely used by minicomputers and by the more powerful personal computers, especially workstations. UNIX was introduced in 1971 and has become widely popular among scientists and others. It is a true multitasking, multiuser system, but was initially intended to be used by programmers and was less "friendly" than some other operating systems. Various versions of UNIX have appeared, and there has been considerable lack of standardization. One version, popular with serious users of personal computers, is called Linux.

unstable A substance that may change spontaneously.

unstable nucleus A nucleus that is not stable and tends to decay to one or more stable nuclei by the spontaneous emission of a charged particle.

vacuum A region devoid of matter; it does not actually exist. In practice it is considered to be a region of very low pressure.

vacuum tubes Electronic devices consisting of an evacuated, usually glass, container, with a heated cathode, an anode, held at a high positive voltage, and various interposed grids and other electrodes capable of exerting a controlling effect on the passage of the current between the cathode and the anode. Vacuum tube diodes, triodes, tetrodes, pentodes, etc. were made in hundreds of different varieties and powers and, prior to the invention of the transistor, were the basis of all electronics and all radio and television transmission and reception. Almost all are now obsolete for such purposes, but vacuum tubes persist in the form of cathode-ray tubes, X-ray tubes, and a few other devices.

valence or **valency** The valency of an element (atom ion or group) is the number of hydrogen atoms (or equivalent) that one atom of the element can combine with or displace. An ion's valence is equal to its charge.

valve A device controlling the passage of fluid in a pipe in one direction.

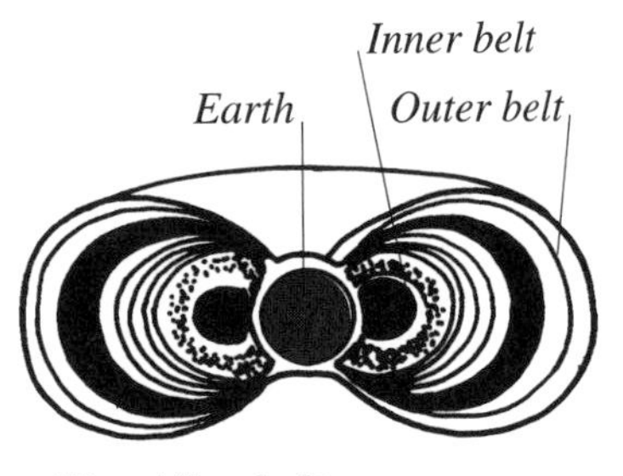

Van Allen belts

Van Allen belts Two zones of high-energy charged particles lying at distances of 620–3,100 miles (1,000–5,000 km) and 9,300–16,000 miles (15,000–25,000 km). The Van Allen belts, named after their discoverer, American physicist James Van Allen, are trapped in the Earth's magnetic field and are sources of intense radiation. The outer belt consists mainly of electrons; the inner also contains protons. They were detected by satellite studies in 1958.

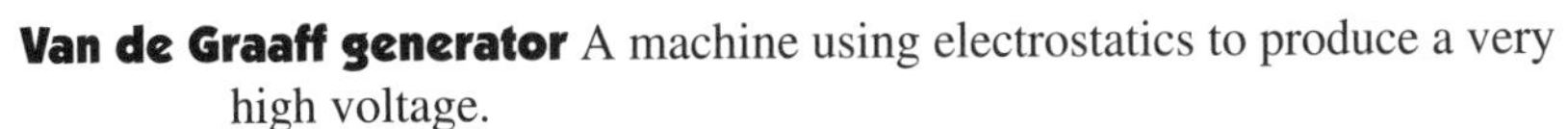

Van de Graaff generator A machine using electrostatics to produce a very high voltage.

Van der Waals forces Weak intermolecular or interatomic forces between neutral molecules or atoms. They are much weaker than chemical bonds.

vapor Gas that is below the temperature at which it can be liquefied by pressure (the critical temperature).

vaporization The process of change of state of a solid or liquid to a vapor.

vapor lamps Light sources in which ionization of a gas forms a conducting path between two electrodes so that a current can pass to excite the gas to emit either mainly visible or ultraviolet light. The gaseous element varies and can be neon, argon, mercury vapor, xenon, sodium metal-halide vapor, or others. Also known as gas discharge lamps.

vapor pressure The pressure exerted by the free molecules or atoms given off by all solids and liquids. If this evaporation is occurring in an enclosed space, an equilibrium between the vapor and the material will be reached that depends on the volatility of the material and the temperature.

VDU *See* visual display unit.

vector A quantity that has both size and direction.

velocity Speed in a specified direction, a vector quantity.

Vernier scale A small graduated scale that moves over a main fixed scale to obtain a more precise value of measurement. If the main scale can give a measurement to one decimal place (given by the point at which the start of the Vernier touches the main scale), the Vernier scale allows a reading to the second decimal place by reading that division on the Vernier that coincides with a reading on the main scale.

very high frequency (VHF) Radio or television frequencies in the range 30 million cycles per second (30 MHz) to 300 million cycles per second (300 Mhz) and having a corresponding wavelength of 1–10 m. The term is a victim of technological progress; such frequencies are no longer considered to be very high.

vibration *See* oscillation.

video amplifier An electronic high-frequency amplifier capable of amplifying, with minimal distortion, alternating voltages of frequencies between about 2–100 million cycles per second (megahertz or Mhz). They are used in TELEVISION receivers, VDUs, and other devices.

virtual image A virtual image is produced by diverging rays so that it appears to be in a particular place, but the rays do not pass through it and it cannot be produced on a screen.

virtual memory A computer arrangement in which the workspace required is larger than the available RAM and some of the material in use is temporarily transferred to an alternative memory space such as a hard disk. When the program refers to a memory location, the

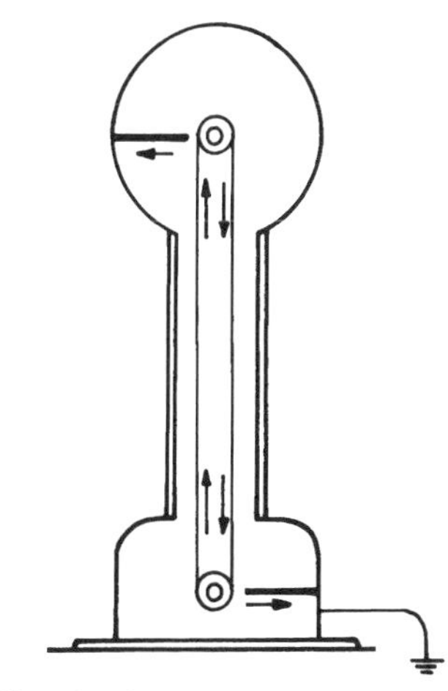

Van de Graaff generator

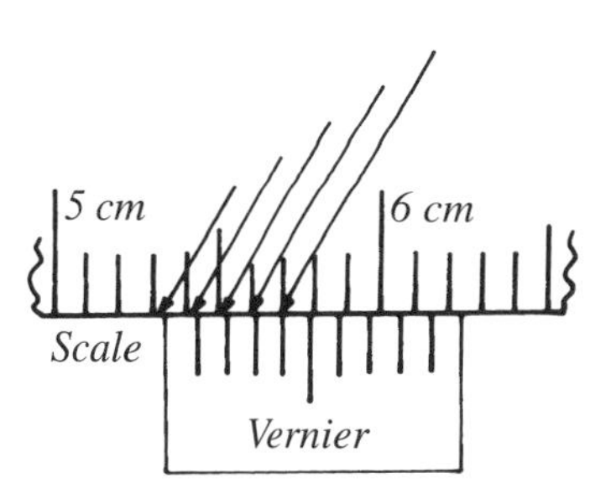

Vernier scale

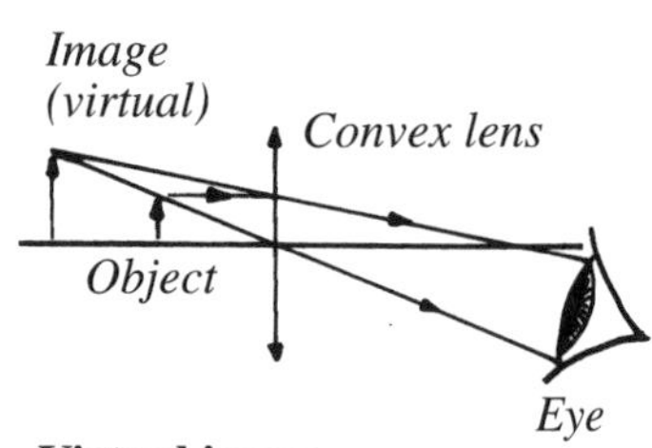

Virtual image
(*See* previous page)

machine checks whether this is in RAM. If it is not, the system temporarily retrieves the needed data from the backing store, making room for it, if necessary, by moving other data to the hard disk or elsewhere.

virtual ray A virtual ray is one that appears to come from a virtual image but does not actually exist.

viscosity A measure of a fluid's unwillingness to flow. The more viscous a fluid, the less easily it flows.

visual display unit (VDU) A computer output device that displays information, either in alphanumeric or graphical form, either currently held in a volatile memory or in files on a disk drive. Most VDUs employ a color cathode-ray tube, but flat, liquid-crystal color displays are becoming more common.

volatile memory A memory whose contents are irretrievably lost if the power supply is switched off. Some volatile memories have a battery power supply backup. In most cases, however, the user will copy the contents to a magnetic disk (backup) before switching off the power.

volt The SI unit of potential difference, electric potential, and electromotive force. It is the potential difference between two points in an electric circuit when a charge of one coulomb moves between the two points, producing one joule of energy.

voltage The value of the potential difference, or electromotive force, expressed in volts.

voltage divider A tapped fixed or variable resistor connected across the full voltage supply. Any lesser voltage, determined by the ratio of the two resistances on either side of the tap, can be obtained between one end of the resistor and the tap.

voltage multiplier A rectifier circuit arranged so that the DC output voltage is equal to two or more times the peak AC input voltage. This is achieved by an arrangement of rectifier diodes and capacitors allowing voltages to be "stacked" one on top of the other. No transformer is necessary, and the system is useful when high-voltage, low-current supplies are needed.

voltage regulation The maintenance of the voltage of a power supply within certain limits, in spite of large changes in the load. Regulator circuits necessarily involve greater power losses than nonregulated power supplies. A reference voltage is provided by a constant voltage diode

(Zener diode), and this is used to maintain the voltage output of a power transistor that acts as a variable-resistance device operating across the output of the power supply, compensating for the effect of changes in the load resistance. For low-power circuits, the Zener diode alone may be sufficient.

voltameter An instrument used to measure electric current. It is an electrolytic cell. A measure of the current is given by the increased mass of the cathode.

volume The amount of space occupied by a body or substance.

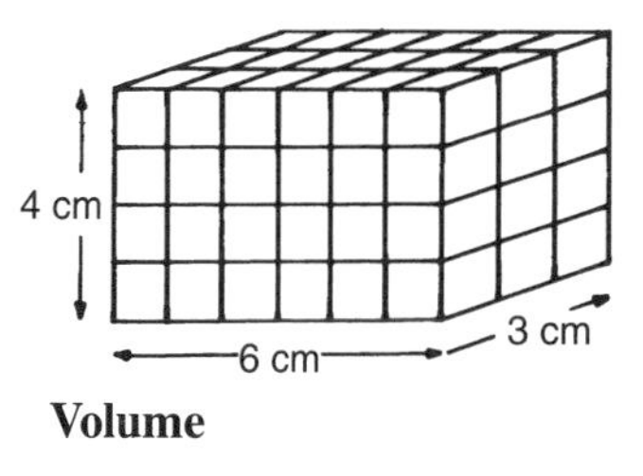

Volume

water cycle The continual movement of water between land, sea, and sky.

water jacket A pipe that fits around another pipe in which a substance requiring to be cooled flows. Water flows in the outer pipe, the water jacket, cooling the substance in the inner pipe.

water of crystalization The exact number of water molecules that are chemically bonded to a molecule of a salt within a hydrated crystalline compound.

watt The SI unit of power, which is one joule per second.

wave A way in which energy is transferred through a medium as a series of periodic oscillations. These involve either the disturbance of particles, as in sound waves, or a periodic change in a physical quantity, as in electromagnetic waves. Waves can be transverse or longitudinal.

waveform The shape of a wave.

wavelength The distance between two corresponding points on a wave – two peaks or two troughs. If the velocity of a wave transmission through a given medium is constant, as is commonly the case, the wavelength will be a reciprocal of the frequency, and vice versa.

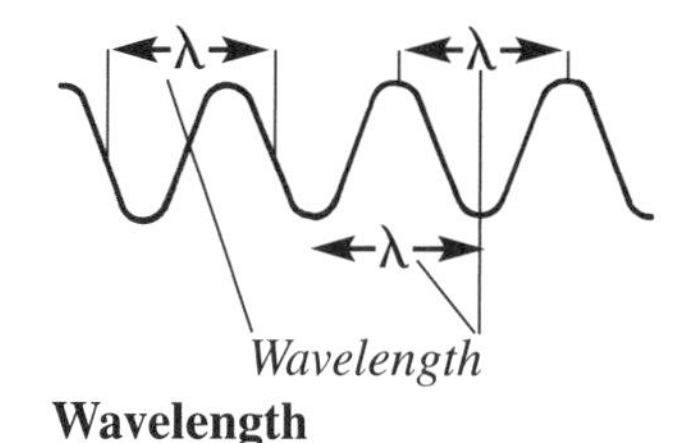

Wavelength

wave mechanics The theory of matter, first postulated by French physicist Louis de Broglie, that elementary particles, such as electrons, protons, and neutrons, can behave as waves. In 1924 de Broglie suggested that they showed the same wave-particle duality then known to be the case with light. This was subsequently proved and became the basis for quantum mechanics.

wavemeter A device for determining the frequency, i.e., the number of cycles per second (Hz), of an electromagnetic wave. Wavemeters work by applying the wave to a system that can resonate at the wave frequency and then detecting the energy produced. At up to about 100 million Hz a variable tuned circuit consisting of an inductance in parallel with a capacitance can be used. Resonance is indicated by a peak in the current flowing in the circuit. Above such frequencies, short wires or wave-guides, adjusted in physical size to match the wavelength being measured, are used.

wave optics A branch of optics that recognizes the wave nature of light and is concerned with such phenomena as diffraction, interference, and polarization. *Compare* geometrical optics.

wave shaping The modification of time-varying electrical current or voltages by passive filters (high-pass, low-pass, or band-pass) or current limiters, or by active electronic means, such as sine-wave, square-wave or sawtooth oscillators, or by various gating circuits.

weight The force exerted on a body by the Earth's gravitational pull. This force is directly proportional to the mass of the body and is affected by its distance from the center of the Earth, being greater at the poles than at the equator by a factor of about 0.3% because the poles are nearer to the center than the equator. The gravitational attraction between two bodies is also inversely proportional to the square of the distance between their centers. The mass of a body is not affected by its relationship to other bodies.

weightlessness An object has no weight if there is no resultant force between it and its surroundings, such as a body in free fall.

Wheatstone bridge An apparatus for measuring electrical resistances.

white dwarf As a red giant star cools and contracts, its core becomes a very small, dense white-hot remnant, a white dwarf star.

white noise Sustained noise of even amplitude containing the whole range of audible frequencies at substantially equal energies for all frequencies. The hiss of escaping steam resembles white noise. When certain bands of frequency predominate, the noise may be described as "pink noise," "blue noise," and so on.

work The transfer of energy occurring when a force is applied to a body to cause it to move. The work done is the product of the force and the distance moved by its point of application along the line of action of the force. Work can be considered as one manifestation of energy.

work function (electron affinity) The minimum energy required to remove an electron from a material to allow it to exist outside the material in a vacuum. The thermionic work function is the application of heat to remove an electron. The photoelectric work function is the application of high-energy radiation to remove an electron.

workstation A powerful personal computer with a large RAM, usually a reduced instruction set (RISC) high-speed microprocessor, a Unix multitasking operating system, a 32-or 64-bit architecture, and a substantial disk storage system. Workstations are capable of carrying out work that, two or three decades ago, would have required a mainframe or minicomputer.

X-ray astronomy The study of the X rays produced by sources outside the solar system.

X rays The electromagnetic radiation produced when a high-speed beam of electrons, accelerated by a high voltage, strikes a metal, such as copper or tungsten. This radiation, which comes from the tightly bound inner-shell electrons of atoms, is of a frequency between that of ultraviolet light and that of gamma rays and has the power to penetrate matter to varying degrees, depending on the shortness of its wavelength and the density of the matter. X rays act on photographic film in much the same way as does visible light. These properties make x-radiation extremely valuable in scientific research, industry, and medicine, and X rays have been in use for diagnostic purposes for more than a century. X rays have wavelengths of between 10^{-12} and 10^{-9}.

yield point The point (yield stress) at which a material under increasing stress undergoes substantial irreversible deformation.

Young's double-slit experiment This demonstrates the interference of light waves. The slits give point sources of light with the same brightness and frequency. A pattern of light and dark lines (Young's fringes) is produced by the constructive and destructive interference between the two sources.

Young's modulus The ratio of any stress (force) applied to an elastic body to the resulting strain (deformation), so long as the elastic limits are not exceeded. The modulus applies to a tensile or compressive stress applied to such things as a thin rod or to a beam to cause it to bend. It is a measure of the stiffness of a material. The modulus is named after English physicist Thomas Young.

Zeeman effect The splitting of spectrum lines by a strong magnetic field, which affects the magnetic moments of the components of the atomic energy levels.

Zener diode A simple semiconductor device (p-n junction diode) that, when a voltage higher than a stated value is applied to it through a resistance, will immediately pass enough reverse current to reduce the voltage across it to the stated value (the Zener voltage). This diode voltage is maintained over a wide range of applied voltages and can be used as a reference voltage in voltage regulation. The diode was developed by American physicist C. M. Zener.

zero potentiometer *See* potentiometer.

zeroth law A fundamental principle behind the first, second, and third laws of thermodynamics. If two systems are separately in thermal equilibrium with a third, they are in thermal equilibrium with each other.

zone refining A repetitive purifying process using induction heating or other means to cause a narrow molten zone to pass slowly along the specimen. Impurities either raise or lower the melting point and are thus segregated, usually in the same direction as the movement of the zone (because most impurities lower the melting point). The process is repeated until the desired degree of purity is achieved.

SECTION TWO
BIOGRAPHIES

Abbe, Ernst (1840–1905) German physicist who specialized in optical lenses and made many important improvements. Abbe turned an empirical art into a science and, incidentally, made a fortune for Carl Zeiss, with whom he collaborated.

Aiken, Howard Hathaway (1900–73) American applied mathematician and computer pioneer who designed an electromechanical calculator using relays that was completed in 1943. It weighed nearly 5 tons (5,000 kg), was 50 feet (155 m) long, and was obsolete before it was finished. It worked for 16 years, however, calculating mathematical tables.

Airy, Sir George Biddell (1801–92) English astronomer and geophysicist who in the 1850s laid a basis for the theory of isostasy – the state of balance in the Earth's crust where continents of light material float on a denser substance into which deep continental "roots" project like the underwater mass of floating icebergs.

Luis Walter Alvarez

Alhazen (c. 965–c. 1040) Arab physicist and mathematician who specialized in optics, and wrote an influential text covering spherical and parabolic mirrors.

Alvarez, Luis Walter (1911–88) American atomic physicist who greatly advanced the field of high particle physics and, in 1946, developed the proton linear accelerator. He won the Nobel Prize in physics in 1968. Alvarez also proposed, along with his son, geologist Walter Alvarez, Frank Asaro, and Helen Michel, the theory that the extinction of the dinosaurs was caused by the impact of a large meteorite. Their evidence included the discovery of a larger than normal layer of iridium dated to the appropriate time. This is characteristic of large meteorite impacts.

André Marie Ampère

Ampère, André Marie (1775–1836) French physicist and mathematician who gave his name to the basic unit of electric current (ampere, amp). He laid the foundations of the science of electrodynamics (now electromagnetism) through theoretical and experimental work after the discovery by Hans Christian Ørsted in 1820 of the magnetic effects of electric currents.

Anderson, Carl David (1905–91) American physicist who discovered a particle, the positron, with the mass of an electron but with a positive charge. This demonstrated the existence of antimatter. If a positron meets an electron, both are annihilated. Anderson also discovered muons and shared the 1936 Nobel Prize in physics with Victor Hess.

Archimedes (c. 287–212 BCE) Greek mathematician and technologist who found formulas for the volume of a wide range of regular solids and for the area of a range of plane figures. His methods were similar to those of calculus. He originated the science of hydrostatics and discovered that a floating body displaced its own weight of water. He invented the Archimedian screw for raising water and a number of large military weapons. He was the leading figure in rigorous scientific and mathematical thought of the ancient world.

Archimedes

Armstrong, Edwin (1890–1954) U.S. electrical engineer and professor at Columbia University whose inventions in electronics revolutionized radio reception. He showed how radio receiving circuits could be greatly increased in sensitivity by applying positive feedback, and his superheterodyne circuit became the standard for all radios.

Aston, Francis (1877–1945) English atomic physicist who worked with J. J. Thomson at the Cavendish Laboratory, Cambridge. His principal field of study was in elements of equal atomic number but different atomic weight (isotopes). He also invented the mass spectrograph. He was awarded the Nobel Prize in chemistry in 1922.

Auger, Pierre (1899–1993) French physicist who showed that atoms could lose energy in ways other than by emitting radiation. Auger electrons are electrons lost by atoms that change their energy state in this way. Unlike beta particles, they all have identical energy, so are useful in calibrating nuclear detection devices. Auger also described "Auger showers." These are showers of high-energy secondary particles produced by cosmic rays impinging on upper atmosphere atoms.

Baade, (Wilhelm Heinrich) Walter (1893–1960) German-born American astronomer whose work, involving new ways of identifying and classifying stars, led him to increase and improve Edwin Hubble's values for the size and age of the universe. He also worked on supernovae and on radio stars.

Charles Babbage

Babbage, Charles (1791–1871) English inventor who, supported by the mathematician and first computer programer, Ada, Countess Lovelace, designed two large mechanical calculating engines, the second of which could be provided with a program. Neither was finished in his lifetime, but these machines must be regarded as the first full computers.

Babinet, Jacques (1794–1872) French physicist who first suggested that more accurate standards of the measurement of length could be achieved by using the wavelengths of a line in the light spectrum as standards. He proposed that the angstrom unit should be defined in terms of the wavelength of the red line in the spectrum of cadmium.

Francis Bacon

Bacon, Francis (1561–1626) English philosopher and essayist whose book *The Advancement of Learning* (1605) drew serious attention for the first time to the fact that the real source of scientific knowledge was not the authority of pundits such as Aristotle but observation, experimentation, direct experience, and careful induction. This was the start of the scientific method that was to prove so fruitful.

Bacon, Roger (c. 1214–92) English philosopher who tried to compile an encyclopedia containing all the knowledge of his day. The attempt failed but contained much valuable mathematical information and included some remarkable speculations about mechanical transport, heavier-than-air flying machines and the possibility of circling the globe.

Baird, John Logie (1888–1946) Scottish electrical engineer and inventor who was the first to demonstrate working television. Baird used a mechanical system with a large rotating disc with a spiral of slots, allowing only 30 lines per screen, but the system worked and showed that TV was feasible. It was not until the cathode-ray tube was used with an electronic scanning system that reasonable high-resolution TV of entertainment quality was achieved, in 1937.

Bardeen, John (1908–91) American physicist who, with Shockley and Brattain, invented the point-contact transistor in 1947, thereby changing the face of technology and making possible the later development of the personal computer. They were awarded the Nobel Prize in physics in 1956. Bardeen proceeded to develop a convincing theory of superconductivity, which won him a second Nobel Prize, in 1972, which he shared with Leon Cooper and J. Robert Schrieffer.

Barkenhausen, Heinrich Georg (1881–1956) German physicist who discovered that if a slowly and smoothly increasing magnetic field was applied to ferromagnetic material, the magnetization did not increase smoothly but in jumps. This was explained by magnetic domains suddenly changing direction. Barkenhausen also developed an ultrahigh-frequency oscillator that initiated microwave technology.

Barkla, Charles Glover (1887–1944) British physicist who studied the scattering of X rays by various materials and who demonstrated that X rays had the characteristics both of waves and of particles. He also demonstrated X-ray fluorescence and was awarded the 1917 Nobel Prize in physics. Barkla turned into a rigidly minded recluse who cited only his own publications and proposed bizarre theories.

Charles Glover Barkla

Barnard, Joseph Edwin (1870–1949) English physicist and microscopist who pioneered the application of photography to microscopic observations. He also improved the resolving power of the optical microscope by using ultraviolet light. This work was, however, overtaken by the development of the electron microscope.

Bartholin, Erasmus (1625–98) Danish physicist and physician who discovered that a double image was perceived when an object was viewed through Iceland felspar (calcite). This is known as double refraction. The observation led later to an understanding of the polarization of light.

Basov, Nikolai Gennediyevich (b. 1922) Russian physicist who, independently of Townes, invented the maser that used ammonia molecules and stimulated them into uniform high-frequency movement. Using similar principles, he then proceeded to develop the laser and the pulsed laser and worked on semiconductor lasers, without which CDs and CD-ROMs and many other devices would not have been practicable. Basov and Townes shared the 1964 Nobel Prize in physics with Aleksandr Prokhorov.

Bates, Leslie Fleetwood (1897–1978) English physicist who worked on the magnetic properties of materials and who, in 1938, published a successful textbook, *Modern Magnetism.* During World War II, he was concerned with the degaussing (protecting) of ships as a defense against magnetic mines.

Antoine Henri Becquerel

Becquerel, Antoine Henri (1852–1908) French physicist who was the first to discover radioactivity. In 1896 he noticed that a uranium salt laid on a totally enclosed photographic plate caused the plate to be exposed. Becquerel concluded that the salt was emitting rays similar to X rays, which had been discovered by Roentgen the year before. He then studied and described the properties of the natural radioactivity of uranium. For this work, he shared the Nobel Prize in physics with Marie and Pierre Curie in 1903.

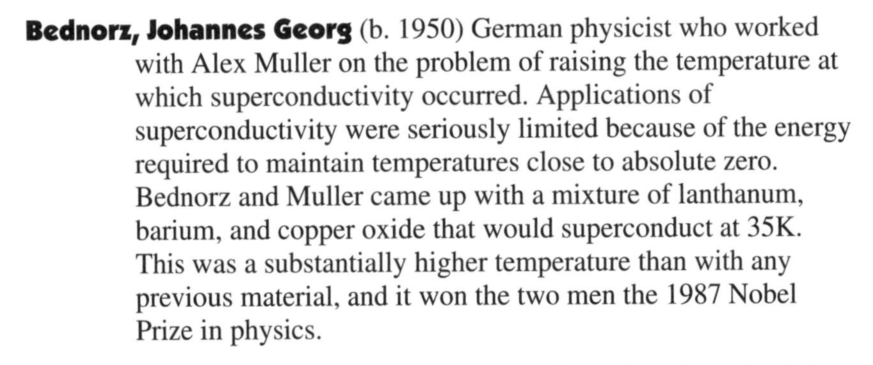

Bednorz, Johannes Georg (b. 1950) German physicist who worked with Alex Muller on the problem of raising the temperature at which superconductivity occurred. Applications of superconductivity were seriously limited because of the energy required to maintain temperatures close to absolute zero. Bednorz and Muller came up with a mixture of lanthanum, barium, and copper oxide that would superconduct at 35K. This was a substantially higher temperature than with any previous material, and it won the two men the 1987 Nobel Prize in physics.

von Békésy, Georg (1899–1972) Hungarian-born American physicist who showed that the then current theory of inner ear cochlear function was incorrect. The basilar membrane fibers did not vibrate in sympathy with particular sound frequencies, like the strings of an undamped piano. Instead, incoming sound causes a traveling wave to sweep along the basilar membrane, and this wave resonates at a peak in a particular region of the membrane. This finding won Békésy the 1961 Nobel Prize in physiology or medicine.

Bell, Alexander Graham (1847–1922) Scottish-born U.S. speech therapist, best known for his invention of the telephone. His interest in the problems of deaf people led him to make many experiments to try to reproduce the human voice electrically. He was convinced of the possibility that sound vibrations in the air could be changed to electrical fluctuations that could pass along a wire and then be reconverted to sound vibrations. He was right.

Alexander Graham Bell

Bell, John Stuart (b. 1928) British physicist who startled the quantum world by describing a possible way of demonstrating, by a statistical analysis of the three coordinates of the spin of a particle, whether or not instantaneous communication between particles, formerly related but now separated by a distance, existed. This work led to an experiment by Alain Aspect, the "Aspect experiment" of 1982, which seemed to suggest that such particles could communicate at speeds greater than the speed of light.

Bernal, John Desmond (1901–71) Irish physicist and crystallographer whose work on the X-ray examination of amino acids, proteins, sterols, and other molecules helped to pioneer modern molecular biology.

Bernoulli, Daniel (1700–82) Dutch-born Swiss mathematician, often known as the father of mathematical physics. Bernoulli's family produced many celebrated mathematicians but he was the most famous. He is best known for his work on trigonometrical functions, continued fractions, hydrodynamics, and the kinetic theory of gases. His famous treatise *Hydrodynamica* was published in 1738.

Jacques Bernouilli

Bernouilli, Jacques or **Jakob** (1654–1705) Swiss mathematician remembered for the Bernouilli numbers – the coefficients found in exponential series – and Bernouilli's theorem in probability theory. He introduced the term *integral* and applied calculus to the study of curves, especially the logarithmic spiral. His analysis of the catenary – the curve formed by a nonelastic string when supported at both ends – has been used in the design of bridges.

Hans Albrecht Bethe

Bethe, Hans Albrecht (b. 1906) German-born American physicist who, on being invited to give a talk on astrophysics, provided details of the only appropriate nuclear reaction (a cyclical series of nuclear reactions starting and ending with carbon-12) that would provide the correct rate of energy production in stars, taking into account the masses of stars. For this seemingly impromptu contribution, Bethe was awarded the 1967 Nobel Prize in physics.

Binnig, Gerd (b. 1947) German physicist who developed the scanning tunneling microscope that can resolve surface molecules and even individual atoms. This work earned Binnig and his colleague, Heinrich Rohrer, the 1986 Nobel Prize in physics, which they shared with Ernst Ruska.

Biot, Jean-Baptiste (1774–1862) French mathematical physicist and astronomer who worked out the basic laws of the rotation of plane-polarized light by optically active crystals and solutions. This work had later value in helping to determine molecular shape. He also studied magnetism at high altitudes while up in a balloon with Gay-Lussac, worked on the refractive indices of gases and the distribution of electric charges on the surface of near-spherical bodies.

Jacob Aall Bonnevie Bjerknes

Bjerknes, Jacob Aall Bonnevie (1897–1975) Norwegian-born American meteorologist, son of Norwegian physicist, Vilhelm Bjerknes. With his father, he formulated the theory of cyclones on which modern weather forecasting is based.

Vilhelm Bjerknes

Bjerknes, Vilhelm (1862–1951) Norwegian mathematician and geophysicist who applied hydrodynamics and thermodynamics to his study of weather systems and produced equations for the thermal energy and other characteristics of developing cyclones.

Blackett, Patrick Maynard Stuart, Baron (1897–1974) English physicist who used the Wilson cloud chamber to show the ejection of a proton from the oxygen isotope newly formed after the alpha particle bombardment of nitrogen. He demonstrated this eight times in 23,000 trials. Blackett improved the Wilson chamber so as to study cosmic rays more economically. He also showed, by studies of terrestrial magnetism, that there had been continental drift in the past 500 million years. He was awarded the Nobel Prize in physics in 1948.

Bloch, Felix (1905–83) Swiss-born American physicist who, in his doctoral dissertation, showed how metals conduct electricity by solving Schrödinger's equation for the movement of electrons in a regular lattice of positive ions. Much of solid-state physics is now based on this work. Bloch also developed nuclear magnetic resonance systems now used in medical, scientific, and industrial scanners. He was awarded the Nobel Prize in physics in 1952 with Purcell for the latter work.

Bohm, David Joseph (b. 1917) American physicist and critic of the indeterminable aspects of quantum theory who tried to suggest that underlying the theory was a more complete and deterministic reality. Bohm worked on the atom bomb project under Oppenheimer and later was persecuted by the Un-American Activities Committee because he refused to give evidence against his friends. He became unemployable and had to leave the country.

Niels Henrik David Bohr

Bohr, Niels Henrik David (1885–1962) Danish physicist who worked with J. J. Thomson at Cambridge and Ernest Rutherford at Manchester. He greatly extended the theory of atomic structure when he explained the spectrum of hydrogen by means of his atomic model and quantum theory. Bohr's contribution to nuclear physics was of the greatest importance, and this is reflected in his award of the 1922 Nobel Prize in physics.

Ludwig Boltzmann

Boltzmann, Ludwig (1844–1906) Austrian theoretical physicist who applied statistical mathematical methods to thermodynamics, thereby greatly advancing understanding. His work was based on the kinetic theory of gases and was little appreciated in his lifetime, but was later seen to be of the greatest importance. Boltzman, a depressive, committed suicide.

Bondi, Sir Hermann (b. 1919) Austrian-born British mathematical physicist who researched radar during World War II and became one of the most outstanding scientific administrators in postwar Britain, concerned with space science, defense, energy, and the environment. He wrote a germinal work on cosmology, but seems to have backed the wrong horse by supporting the "steady state" theory of the universe instead of the "big bang" theory.

Boole, George (1815–64) Self-taught English mathematician and teacher whose work on symbolic logic showed that it was possible to construct an arithmetic using only the number 1 and 0. A century later this idea was to bear remarkable fruit in the design of digital computers, all of which are based on Boole's ideas. He also pioneered the calculus of operators and wrote a book on the calculus of finite differences (1860), which remains a classic.

Boot, Henry Albert Howard (b. 1917) English physicist who worked on the generation of centimetric microwaves for radar systems. Existing oscillators could not produce high-power outputs at wavelengths of a few centimeters, so, in 1939, Boot and his colleague, John Randall, produced the cavity magnetron in which electrons could circulate at extremely high frequencies so as to produce electromagnetic radiation. This device was of vital importance in World War II. The successors to Boot's cavity magnetron are now to be found in the kitchens of most Western homes.

Giovanni Alfonso Borelli

Borelli, Giovanni Alfonso (1608–79) Italian mathematician, physicist, and physiologist who was the first to show how muscular contraction maintains posture and brings about bodily movements. He concluded correctly that muscles are made to contract by an influence that passed from the brain to the muscle by way of a nerve. His book *De Motu animalium* also explained the mechanism of breathing.

Max Born

Born, Max (1882–1970) German physicist whose theory of the quantum mechanics of the motion of an atomic particle showed that quantum mechanics allowed only a statistical interpretation of events at the atomic level. He elucidated the wave/particle paradox by showing that the square of the amplitude of the wave function at a particular point is the probability of finding a particle at that point. He shared the Nobel Prize in physics in 1954 with Walther Bothe.

Bose, Sir Jagadis Chandra (1858–1937) Indian physicist and botanist whose main work was in the polarization and reflection of electric waves and in the investigation of the growth and sensitivity of plants.

Bose, Satyendra Nath (1894–1974) Indian physicist who in 1924, without any reference to classical electrodynamics, succeeded in deriving the black-body radiation law of Planck. Planck had used classical electrodynamics. Einstein followed up Bose's work to produce a system of quantum mechanics for integral spin particles, now called Bose-Einstein statistics. The boson particle is named after Bose.

Robert Boyle

Boyle, Robert (1627–91) Irish physicist and chemist, and cofounder of the Royal Society, whose book *Sceptical Chymist* (1661) defines the chemical element as the practical limit of chemical analysis. The celebrated Boyle's law (1662) states that, the temperature being kept constant, the pressure and volume of a gas are inversely proportional.

Bragg, Sir William Henry (1862–1942) English physicist who, working with his son, used X-ray diffraction methods to determine crystal atomic structure, thereby founding what amounted to a new science: the discipline of X-ray crystallography, by which it was possible to determine the structure of crystals by X-ray diffraction. He shared the 1915 Nobel Prize in physics with his 25-year-old son, William L. Bragg, a unique event.

Bragg, Sir William Lawrence (1890–1971) British physicist and X-ray crystallographer who worked with his father in the development of crystallography and was the youngest Nobel Prize winner. Later, as head of the Cavendish Laboratory in Cambridge, England, Bragg encouraged Watson and Crick in their search for the structure of DNA using X-ray crystallographic techniques. The celebrated *Nature* paper on DNA was sent to the editor with Bragg's recommendation and was published quickly.

Walter Brattain

Brattain, Walter (1902–87) American physicist who, working with Bardeen and Shockley at Bell Laboratories, invented the point-contact transistor. This was the first electronic amplifying and switching device that required only a single, low-voltage power supply and was the precursor of the junction transistor. It changed the world. Brattain shared the 1956 Nobel Prize in physics with Bardeen and Shockley.

Brewster, Sir David (1781–1868) Scottish physicist, interested in the study of optics, who showed that light could be plane-polarized by shining a beam on a plane glass surface at a particular angle. He also showed that the refractive index of the glass could be obtained from the measurement of this angle.

Percy Williams Bridgman

Bridgman, Percy Williams (1882–1961) U.S. physicist noted for his work on the effects of very high pressures on matter. He showed that most liquids became more viscous under such pressures. He was awarded the Nobel Prize in physics in 1946.

Brockhouse, Bertram N. (b. 1918) Canadian physicist who, working independently of Clifford Shull, developed the new technique of neutron scattering analysis. Beams of neutrons are thrown on the nuclei of the atoms of material to diffract and produce an information-loaded pattern. The method allows the investigation of the atomic structure of bulk matter and promises to have numerous important practical applications. It earned Brockhouse and Shull the 1994 Nobel Prize in physics.

de Broglie, Louis Victor Pierre Raymond Prince (later Duke) (1892–1987) French physicist who was the first to suggest the unity of waves and particles expressly with reference to the electron. He did research on quantum theory, was the founder of wave mechanics, and was awarded the Nobel Prize in physics in 1929.

Robert Wilhelm Bunsen

Bunsen, Robert Wilhelm (1811–99) German experimental chemist and inventor. He developed the gas burner that bears his name and the ice calorimeter. Working with the German physicist Kirchhoff, he developed the important analytical technique of chemical spectroscopy. Bunsen also discovered the elements cesium and rubidium.

Bush, Vannevar (1890–1974) American electrical engineer who worked on electromechanical analog computers and invented the differential analyzer, which could solve differential equations. He also developed a code-breaking machine that was effective against Japanese encoded messages in World War II. Almost all such analog computers have now been superseded by digital machines.

Cabeo, Niccolò (1586–1650) Italian physicist who believed, wrongly, that electromagnetic effects, then known as "effluvium," operated by pushing aside the air around the current conductor. The idea of effluvium dominated electrical theory until the middle of the 18th century.

Cantor, George Ferdinand Ludwig Philipp (1845–1918) Russian-born German mathematician whose work on set theory and the analysis of infinite sets transformed the foundations of mathematics. Cantor's set theory was of great importance, but involved paradoxes. These he recognized and reconciled by faith in a God whose mathematical ability was beyond human understanding.

Cardano, Geronimo (1501–76) Italian mathematician and physician who produced the first printed work on algebra, the *Ars magna* (1545), and described the general method for solving cubic equations. He also published books on medicine, arithmetic, and philosophy and an encyclopedia of inventions and experiments called *De subtilitate rerum* (1550).

Carlson, Chester Floyd (1906–68) American inventor who developed the system of dry photocopying using a selenium-coated electrically charged drum on which the image of the material to be copied was focused by a lens. Light discharges selenium, so a latent image was formed that would attract a fine black carbon powder. This would be transferred to charged paper and fused in by heat. The idea was bought by a firm that became Xerox.

Carnot, Nicolas Leonard Sadi (1796–1832) French military engineer with no formal scientific education who developed the theory of the heat engine and published work that led to the science of thermodynamics. He proved that perpetual motion was impossible and prompted the law of conservation of energy. His work inspired and suggested the direction of thought of many later theoretical physicists. The Carnot cycle is known to all heat engineers.

Nicolas Leonard Sadi Carnot

Edmund Cartwright

Cartwright, Edmund (1743–1823) English inventor who produced a power loom that greatly increased the output possible with looms worked by treadle or hand. He also developed a wool-carding machine. The power loom aroused much opposition but brought prosperity – and the "dark, satanic mills" – to Britain. Cartwright failed as a businessman and went bankrupt, but Parliament voted him £10,000, a fortune at that time.

Carty, John Joseph (1861–1932) American engineer whose technical advances in telephony led to the wide commercial and transcontinental use of the telephone. He developed urban telephone networks that were supplied with electric current from a central source.

Casimir, Hendrik Brugt Gerhard (b. 1909) Dutch physicist who concentrated on low-temperature physics and superconductivity. To account for the close interrelationship of the magnetic and thermal properties of superconductors, he proposed a model of superconductivity in which a proportion of the electrons were superconducting while the rest behaved normally.

Cassegrain, N. (fl. 1672) French telescope designer whose name is given to the design of reflecting telescopes with a small secondary hyperboloid convex mirror that reflects through a hole in the center of the main mirror to an eyepiece.

Cassini, Giovanni Domenico (Jean Dominique) (1625–1712) Italian-born French astronomer who became professor of astronomy at Bologna in 1560 and first director of the observatory at Paris in 1669. He greatly extended knowledge of the Sun's parallax, the periods of Jupiter, Mars, and Venus, zodiacal light, and other astronomical phenomena.

Sir George Cayley

Cauchy, Augustin Louis (1789–1857) French military engineer and pure mathematician who produced 16 named mathematical concepts and theorems and published seven books and more than 700 professional papers. His principal contribution was the correction of defective logic in complex analysis and calculus.

Cayley, Sir George (1773–1857) English aeronautical scientist who devoted his life to the study of heavier-than-air flying machines. He built numerous models and a human-carrying glider, and worked out the function of the tail surfaces in steering, climbing, and diving. His machines might have preempted the success of the Wright brothers had he had a sufficiently powerful and light engine.

Sir James Chadwick

Chadwick, Sir James (1891–1974) British physicist who discovered the neutron, the existence of which he had predicted from his interpretation of an experiment described by the Joliot-Curies. Chadwick then confirmed his opinion by experiment in 1932, thereby making one of the great advances in knowledge of the atom. He was awarded the Nobel Prize in physics in 1935.

Chance, Britton (b. 1913) American biophysicist who proved that enzymes work by attaching themselves to the substance on which they act (the substrate). He achieved this by a spectroscopic technique using the enzyme peroxidase, which contains iron and absorbs certain light wavelengths strongly. Chance also helped to work out the way cells get their energy from sugar, by observing that concentrations of adenosine diphosphate (ADP) were related to the oxidation-reduction states of the proteins in the respiratory chain.

Chandrasekhar, Subramanyan (1910–95) Indian-born American astrophysicist who showed that when the nuclear fuel of a star is exhausted, gravitational forces begin to pull the material together until it is very dense, atomic electrons are stripped off, and outward pressures from the nuclei halt the compression, thus leaving, in most cases, a white dwarf. Chandrasekhar showed, however, that the greater the mass the smaller the radius, and that a star with a mass greater than 1.4 times the mass of the Sun cannot evolve into a white dwarf. This is known as the Chandrasekhar limit. For this and other advances in astrophysics, Chandrasekhar was awarded the 1983 Nobel Prize in physics, with William Fowler.

Jacques Alexandre César Charles

Charles, Jacques Alexandre César (1746–1823) French physicist and physical chemist who, with Gay-Lussac, established a law of the changes in gas volume caused by temperature changes at constant pressure. This is commonly known as Charles' law of pressures.

Cherenkov, Pavel Alekseyevich (1904–90) Russian physicist who shared the 1958 Nobel Prize in physics with Igor Tamm and Ilya Frank for the demonstration that water bombarded with gamma rays emits a blue light, known as the "Cherenkov effect," which has applications in particle identification.

Chu, Steven (b. 1948) American physicist who, working independently of Phillips and Cohen-Tannoudji, developed a method of using laser light to cool and trap atoms, allowing them to be studied in detail. For this work, which has greatly extended knowledge of the relationship between matter and radiation and which promises important practical applications, Chu and Phillips shared the 1997 Nobel Prize in physics with Cohen-Tannoudji.

Clairaut, Alexis Claude (1713–65) French geophysicist who established that the Earth is not spherical but is flattened at the poles – it is an oblate spheroid.

Rudolf Julius Emanuel Clausius

Clausius, Rudolf Julius Emanuel (1822–88) German theoretical physicist who greatly advanced the ideas of Carnot and Joule, thereby largely establishing thermodynamics. He cleared up previous difficulties by pointing out that heat cannot pass spontaneously from a cold to a hot body, and furthered the understanding of the kinetic theory of gases. He also promoted the concept of entropy.

Cockroft, Sir John Douglas (1897–1967) English physicist who showed that atoms could be transmuted by bombardment of their nuclei with particles. Cockcroft, in collaboration with Walton, designed a particle accelerator using high voltages to move protons at high speeds. Cockcroft and Walton shared the Nobel Prize in physics in 1951.

Cohen-Tannoudji, Claude (b. 1933) French physicist who, working independently of American colleagues Chu and Phillips, developed a method of using laser light to cool and trap atoms, allowing them to be studied in detail. For this work, which has greatly extended knowledge of the relationship between matter and radiation and which promises important practical applications, Cohen-Tannoudji, Chu and Phillips were awarded the Nobel Prize in physics in 1997.

Compton, Arthur Holly (1892–1962) American physicist who in 1923 discovered that X rays passing though paraffin wax suffered an increase in wavelength. This indicated that they were behaving as particles as well as waves (the Compton effect). For this discovery, which had notable theoretical implications, he shared the Nobel Prize in physics in 1927 with Charles Wilson.

Conway, John Horton (b. 1937) British mathematician who, in the course of studying the mathematics of population survival under stipulated rules, invented the popular computer game of Life, and showed that there are patterns that behave like self-replicating animals. Conway has also advanced knot theory and group theory.

Coriolis, Gaspard Gustave (1792–1843) French physicist who applied his studies of a spinning surface to such phenomena as the way that weather and ocean current patterns differed in the Northern and Southern Hemispheres. He is remembered for the "Coriolis effect," which describes the force acting on mobile objects on the Earth's surface. He was the first to coin the term *kinetic energy*.

Gaspard Gustave Coriolis

Coster, Dirk (1889–1950) Dutch physicist who, working with the Hungarian-Swedish George de Hevesy in Copenhagen, discovered the element hafnium in 1923.

de Coulomb, Charles Augustin (1736–1806) French physicist best known for his demonstration of the fact that magnetic and electric attraction are both inversely proportional to the square of the distance separating the sources of the fields. Coulomb also worked on viscosity and friction. His name is remembered in the unit of quantity of electricity moved in 1 second with a current of 1 amp (the coulomb).

Charles Augustin de Coulomb

Cowan, Clyde Larrain (1919–74) American physicist who, working with Reines, demonstrated that the neutrino, a particle predicted by Pauli in 1930, actually existed. Neutrinos are thought to have zero rest mass and to travel at the speed of light. They have no charge, so are very hard to detect.

Crookes, Sir William (1832–1919) English chemist and physicist who discovered the element thallium and invented the radiometer.

Another invention that has had a major social effect was the Crookes tube, by means of which Crookes demonstrated cathode rays (streams of electrons) and showed that they could be deflected by magnets. Cathode-ray tubes are now present in television sets and computer monitors in almost every home in the developed world. At age 68 he began investigating radioactivity.

Marie Curie

Curie, Marie (née Sklodowska) (1867–1934) Polish-born French physicist and wife of Pierre Curie, with whom she worked on magnetism and radioactivity, a term she invented in 1898. Their work on radioactivity earned them the 1903 Nobel Prize in physics, shared with Becquerel. Marie Curie isolated polonium and, in 1910, pure radium, and for this work was awarded the Nobel Prize in chemistry in 1911. She died from leukemia, a consequence of long exposure to radioactivity.

Pierre Curie

Curie, Pierre (1859–1906) French physicist, husband of Marie Curie, who in 1880 discovered piezoelectricity, the property of certain crystals to deform slightly in an electric field and to produce such a field if deformed. Without piezoelectricity, personal computers, ultrasound scanners, and cheap phonograph pickups would have been impossible. He also showed that ferromagnetic materials lose their magnetism at certain temperatures (the Curie point). Curie worked with his wife on radioactivity and showed that emitted particles are either electrically negative (beta particles), positive (alpha particles), or neutral (gamma rays). Curie shared the 1903 Nobel Prize in physics with his wife and Becquerel.

Darwin, Sir George Howard (1845–1912) English geophysicist and second son of Charles Darwin. Howard was a mathematician who studied the Earth-Moon system and proposed that the Moon was formerly much nearer the Earth than it is now, and that the effect of the Moon-induced tides has been to slow the Earth's rotation and cause the Moon to recede. This theory is not universally accepted by cosmologists.

Davisson, Clinton Joseph (1881–1958) American physicist who took up research with Bell Telephone and, working with Germer on electron scattering from nickel, observed an electron diffraction pattern as predicted by de Broglie's theory of the identity of waves and particles. This was important in advancing quantum theory. Davisson shared the 1937 Nobel Prize in physics with George Thomson.

Debye, Peter Joseph William (1884–1966) Dutch-born American physicist and physical chemist who studied molecular structures, the distribution of electric charges in molecules, and the distances between atoms (the turning effect of a force). He studied dielectric constants and developed the theory of dipole moments, showing how these could be applied to understanding the three-dimensional shape of simple molecules, such as those of water. He showed that water molecules were bent and that benzene rings are flat. He received the Nobel Prize in chemistry in 1936 and is remembered in the unit of dipole moment, the debye.

Peter Joseph William Debye

De Forest, Lee (1873–1961) U.S. physicist who founded electronics when he put a third electrode, the grid, into a diode thermionic valve. This electrode, lying between the filament (cathode) and the plate (anode), was able, by the imposition of a small voltage, to control the passage of a large current through the valve (the audion or triode). Amplification had been achieved and the radio industry was born.

Dehmelt, Hans Georg (b. 1922) German-born American physicist who succeeded in isolating a single electron in a container using electric and magnetic fields. This won him a share of the 1989 Nobel Prize in physics with Wolfgang Paul and Norman Ramsey.

Delbruck, Max (1906–81) German-born American biophysicist who, while at California Institute of Technology, did much to create bacterial and bacteriophage genetics, and, in 1946, showed that viruses can effect recombination of genetic material.

Dewar, Sir James (1842–1923) Scottish chemist and physicist who invented the Dewar (or vacuum or Thermos) flask, an insulated, double-walled flask with the inner space evacuated of air to limit heat conduction, silvered to reflect and reduce heat loss by radiation, and with a thick cork to limit loss by convection. Working with Sir Frederick Abel, he invented cordite.

Diesel, Rudolf (1858–1913) French-born German engineer who designed a highly efficient heat engine in which the air-fuel mixture is detonated by the temperature produced by high compression rather than by a spark or a flame. Diesel's engine featured greater fuel economy than the gasoline engines of the time and is now in widespread use in transportation.

Diophantus (fl. 3rd century CE) Greek mathematician who is credited with the invention of algebra and who wrote *Arithmetica,* a work of which six of the 13 books survive.

Dirac, Paul Adrien Maurice (1902–84) English mathematical physicist. He worked on quantum mechanics and gave a complete mathematical formulation of Einstein's relativity theory in *The Principles of Quantum Mechanics* (1930). Dirac was one of the most creative theoretical originators at work in the early 20th century. He shared the 1933 Nobel Prize in physics with Schrödinger.

Christian Doppler

Doppler, Christian (1803–53) Austrian physicist whose principal contribution to science was the important principle, applicable to any wave phenomenon, that if the source of the waves is moving relative to the observer, the perceived frequency will be higher as the source approaches and lower as it retreats. The phenomenon is familiar to all in the context of sound, but applies also to light and other electromagnetic waves.

Douglas, Donald Wills (1892–1981) American aeronautical engineer who produced the first aircraft that could lift a cargo exceeding its own weight. This led to the establishment of the Douglas Aircraft Company and the celebrated line of DC- aircraft. DC-3 airplanes, originated in 1935, are still flying.

Eastman, George (1854–1932) American inventor who in 1880 developed the dry photographic plate and film and, in 1884, produced the first photographic roll film. Personal photography first became widely popular when Eastman produced a small, hand-held box camera in 1888. The Eastman Kodak roll-film camera followed, and then the $1 box Brownie. Eastman made a fortune, gave most of it away to educational institutions, and later committed suicide.

Eddington, Sir Arthur Stanley (1882–1944) English astrophysicist whose work on the composition of stars had a major impact on astrophysics and cosmology. He established the law of the relationship of the mass of a star to its luminosity, and showed that the inward gravitational pressures in a star must be in exact balance with the outward forces caused by gas pressure and radiation.

Einstein, Albert (1879–1955) German-Swiss-American mathematical physicist who ranks with such figures as Galileo and Newton as one of the great conceptual revisors of humanity's understanding of the universe. World-famous for his Special (1905) and General (1915) theories of relativity, and for his formulation of the mass-energy equation, $E = mc^2$, which showed the possibility of atomic weapons and power stations. He was awarded the Nobel Prize in physics in 1921.

Elsasser, Walter Maurice (1904–1991) German-born American geophysicist who was the first to propose that the Earth's magnetic field is caused by the electric currents flowing in the molten iron outer part of the core of the Earth. He also studied the slow alterations (secular changes) in the Earth's magnetic field.

Euclid (fl. 300 BCE) Greek mathematician and teacher who wrote the best-selling mathematical text of all time, *The Elements*. This work of plane geometry has been in manuscript copy and print continuously for well over 2,000 years and has delighted (or disgusted) countless millions of schoolchildren. It was the first mathematical book to be printed. Euclid's geometry was generally thought to be the last word on the subject until Riemann proved, in 1854, that there existed a whole class of non-Euclidean geometries, concerned with curved space.

Ewing, William Maurice (1906–74) American physicist and marine geologist who performed extensive geophysical explorations of the oceans. He measured the thickness of the Earth's crust under the oceans and showed it to average 4 miles (6.4 km), compared with 22–25 miles (35–40 km) under continents. He investigated the transmission of sound waves through water and their reflection from the sea bed, and produced data that led eventually to the acceptance of the plate tectonic theory.

William Maurice Ewing

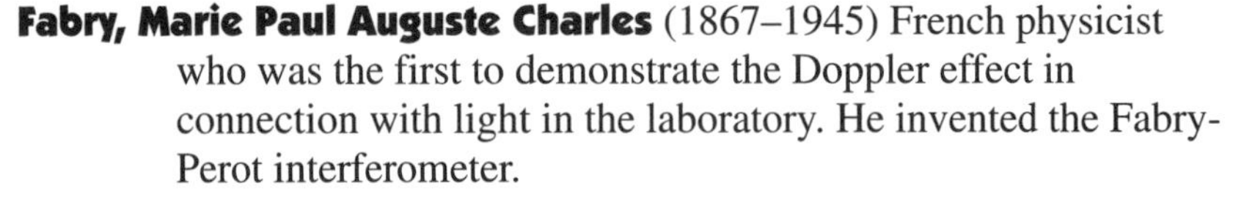

Fabry, Marie Paul Auguste Charles (1867–1945) French physicist who was the first to demonstrate the Doppler effect in connection with light in the laboratory. He invented the Fabry-Perot interferometer.

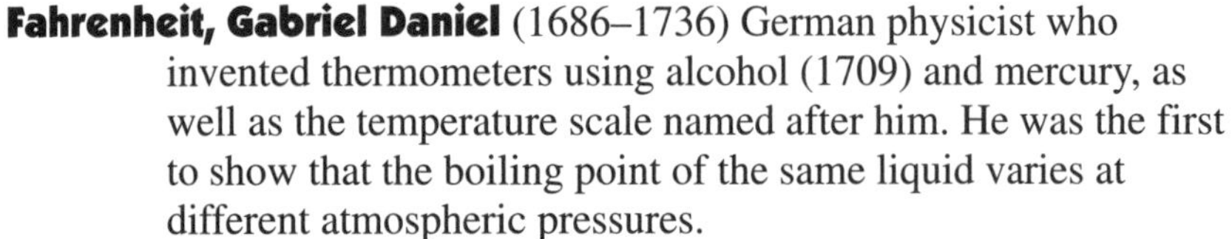

Fahrenheit, Gabriel Daniel (1686–1736) German physicist who invented thermometers using alcohol (1709) and mercury, as well as the temperature scale named after him. He was the first to show that the boiling point of the same liquid varies at different atmospheric pressures.

Michael Faraday

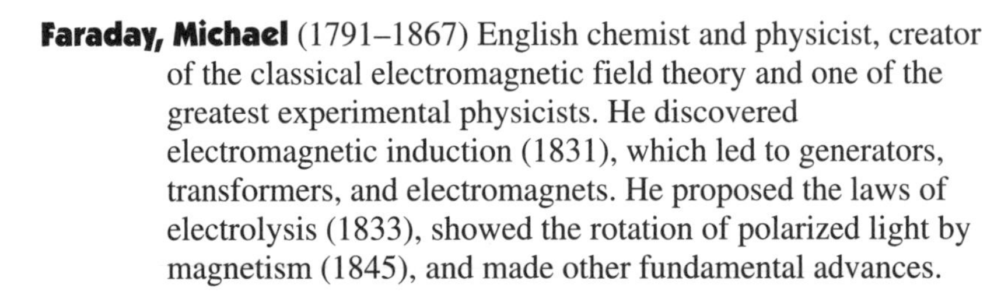

Faraday, Michael (1791–1867) English chemist and physicist, creator of the classical electromagnetic field theory and one of the greatest experimental physicists. He discovered electromagnetic induction (1831), which led to generators, transformers, and electromagnets. He proposed the laws of electrolysis (1833), showed the rotation of polarized light by magnetism (1845), and made other fundamental advances.

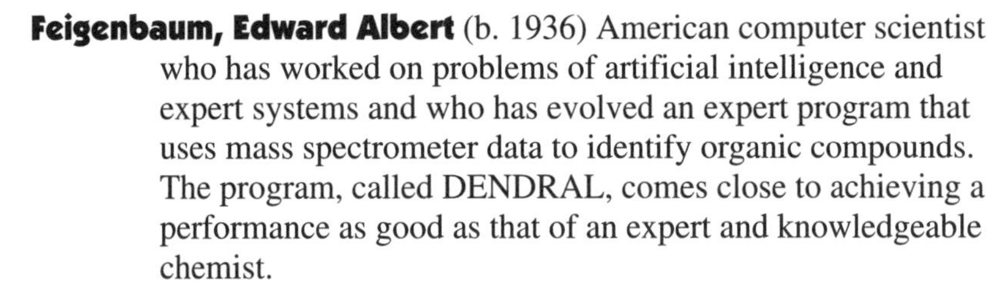

Feigenbaum, Edward Albert (b. 1936) American computer scientist who has worked on problems of artificial intelligence and expert systems and who has evolved an expert program that uses mass spectrometer data to identify organic compounds. The program, called DENDRAL, comes close to achieving a performance as good as that of an expert and knowledgeable chemist.

de Fermat, Pierre (1601–65) French mathematician who contributed notably, among many other things, to the theory of numbers, probability theory, analytic geometry, and geometrical optics. Fermat's last theorem, once the most celebrated unsolved problem in mathematics, states that there are no positive whole numbers x, y, and z, such that $x^n + y^n = z^n$, if n is greater than 2. Fermat had a proof for the case in which n = 4. The general proof was finally provided in 1993 by Andrew Wiles of Princeton University.

Fermi, Enrico (1901–54) Italian-born American physicist who invented quantum statistics (Fermi-Dirac statistics) and discovered particles, now called fermions, which obeyed these principles. He also developed a theory of beta decay that included the yet undiscovered neutrino. He discovered that neutrons slowed by passage through paraffin were much better than fast neutrons in initiating nuclear reactions and developed the first atomic pile. He was awarded the Nobel Prize in physics in 1938.

Feynman, Richard Phillips (1918–88) Colorful American physicist, inspired teacher, and popularizer of science who worked on the application of quantum theory to the interactions between radiation and particles (quantum electrodynamics) using graphical representations that became known as Feynman diagrams. His work on quantum electrodynamics earned him a share in the Nobel Prize in physics in 1965.

Fizeau, Hippolyte Armand Louis (1819–96) French physicist who developed a method of measuring the speed of light using a rapidly rotating toothed wheel that intermittently cut off the beam. He also measured the speed of electrical conduction in a wire.

Fleming, Sir John Ambrose (1849–1945) British physicist and electrical engineer who invented the first vacuum tube, a simple thermionic diode consisting of a heated filament and a positively charged plate. Electrons "boiled" off the filament were attracted to the plate, but current could not pass from plate to filament. The device was thus a rectifier that converted alternating current to direct current.

Jean Bernard Léon Foucault

Foucault, Jean Bernard Léon (1819–68) French physicist who determined the speed of light by a revolving mirror method and proved that light travels more slowly in water than in air (1853). In 1851, using a freely suspended pendulum, he proved that the Earth rotates.

Fourier, Jean-Baptiste Joseph (1768–1830) French mathematician best known for his introduction of the concept that almost any function of a real variable can be expressed as a series of the sines and cosines of integral multiples of the variable. Thus, for example, complex musical tones can be synthesized from sine waves of different multiples of the fundamental frequency. This is called Fourier analysis.

Joseph von Fraunhofer

von Fraunhofer, Joseph (1787–1826) German physicist who founded an optical institute in Munich in 1807 and improved prisms and telescopes, enabling him to discover the dark lines in the Sun's light spectrum, called Fraunhofer lines. He used the lines as reference points for measuring the indices of refraction and the dispersive powers of various samples of glass. He also invented color-fringe-free (achromatic) lenses for telescopes.

Augustin Jean Fresnel

Fresnel, Augustin Jean (1788–1827) French physicist most remembered for a way of constructing uniformly thin but powerful lenses that use stepped glass. He proposed a wave theory of light and predicted changes in its speed through different media. He was interested in polarization, interference, and diffraction of light and wrote a work in three volumes that, at the time of his death, contained almost everything that was known about optics.

Frontinus (d. c. 106 CE) Roman soldier and governor of Britain who was responsible for organizing the water supply to Rome. He wrote a book about aqueducts.

George Gamov

Gamov, George (1904–68) Russian physicist who, using wave mechanics, calculated that protons of quite low energy could penetrate the nuclei of light atoms, and showed that alpha decay could be explained by quantum tunneling. Gamov (and Georges Lemaître) originated the "big bang" theory and asserted that the resulting background radiation must exist throughout the universe. The radiation was later picked up on a radio receiver by Robert Wilson and Arno Penzias.

Gauss, Karl Friedrich (1777–1855) German mathematician, astronomer, and physicist who made numerous important contributions to mathematics, especially in number theory and algebra, and in the geometry of curved surfaces (non-Euclidean geometry). He even proposed the possibility of a non-Euclidean geometry of space, thereby anticipating Einstein, and developed a system of Gaussian coordinates for use on curved surfaces. He also calculated planetary orbits and made contributions to the understanding of the Earth's magnetic field, and optical systems of lenses.

Gay-Lussac, Joseph Louis (1778–1850) French chemist and physicist who investigated the expansion of gases with rising temperature and independently formulated the law known as

Charles' law. He investigated the laws of combination of gases and the properties of many chemical compounds and elements, especially the halogens. He also compiled charts of the solubility of many compounds.

Geiger, Hans Wilhelm (1882–1945) German physicist who, working with Ernest Rutherford in 1908, developed an instrument for detecting alpha particles. This was a gas-filled tube with electrodes held at a high voltage relative to each other. The arrival of an alpha particle caused ionization of the gas so that it became conducting, allowing a spike of current to pass, and causing a deflection of a meter needle. Later, Geiger and Walther Müller improved the device. The modern form also detects beta and gamma rays.

Hans Wilhelm Geiger

Gell-Mann, Murray (b. 1929) American theoretical physicist who studied subatomic particles, explaining the properties of some of them by assigning them a quantum number he called "strangeness," and the two quantum properties of "up" and "down." Gell-Mann introduced the arbitrary term *quark*, of which there were three kinds, for the constituents of elementary particles. The quarks are "up," "down," and "strange," and have fractional electric charges compared to that of the electron. For this work, which enabled him to predict the existence of a particle that was later discovered, he was awarded the Nobel Prize in physics in 1969.

Gilbert, Walter (b. 1932) American physicist, biochemist, and molecular biologist who isolated the repressor molecule that causes genes to be operative or nonoperative, and described the DNA nucleotide sequence to which it binds. For this work, he was awarded the 1980 Nobel Prize in chemistry.

Gilbert, William (1544–1603) English physician whose book on magnetism, *De Magnete* (1600), was the first substantial work on real science to be published in English. Gilbert believed, correctly, that electricity and magnetism are closely related. He distinguished conductors from insulators, showed how to make magnets by hammering an iron bar while it was oriented in the Earth's magnetic field, and he believed that the Earth was a giant magnet.

Glaser, Donald Arthur (b. 1926) American physicist who invented the bubble chamber, which is used to detect the paths of elementary particles. The chamber contains a superheated liquid and the tract of a particle is marked by a line of tiny bubbles. This device, which enabled many new subatomic particles to be detected, earned him the 1960 Nobel Prize in physics.

Kurt Gödel

Gödel, Kurt (1906–1978) Hungarian-born American mathematician best known for his theory of mathematical logic ("Gödel's proof"), which showed that any consistent system capable of describing simple arithmetic will contain propositions that can neither be proved nor disproved by that system. This work led to the branch of mathematics known as proof theory. Gödel also worked on general relativity and cosmology.

von Goethe, Johann Wolfgang (1749–1832) German poet, novelist, lawyer, philosopher, prime minister, physicist, botanist, geologist, and comparative anatomist. Because of his status as an acknowledged genius, Goethe's views on scientific matters commanded a good deal of respect in his time, but his real contribution to the advancement of science was negligible and some of his assertions were seriously misleading. He tried to refute Newton's theory of light and proposed a theory of color vision. Even so, Goethe inspired many people to study science.

Thomas Graham

Graham, Thomas (1805–69) Scottish chemist and physicist. One of the founders of physical chemistry, his research on the molecular diffusion of gases led him to formulate the law "that the diffusion rate of gases is inversely proportional to the square roots of their densities." This is known as Graham's law of diffusion.

Gray, Stephen (1666–1736) English physicist who showed how to make a simple microscope from a drop of water and who, in the course of his experiments with static electricity, demonstrated the conduction of electricity along a string supported by insulating silk threads. The same string supported by brass wire quickly lost its charge.

Gregory, David (1659–1708) Scottish mathematician who wrote books on geometry, astronomy, and optics, and indicated the possibility of designing lenses that would be free of the annoying property of forming colored fringes to images (achromatic lenses).

Gregory, James (1638–75) Scottish mathematician and physicist who designed an improved reflecting telescope, which he described in his book *Optica promota* (1663), and published papers on methods of determining the areas of geometrical figures.

Grimaldi, Francesco Maria (1618–63) Italian physicist who discovered that light was deviated at the edge of an obstacle in its path (optical diffraction) and inferred that light was a wave phenomenon. He made an early lunar map and originated the tradition of naming the Moon's features after famous scientists.

Otto von Guericke

von Guericke, Otto (1602–86) German physicist and engineer who in 1650 invented a vacuum pump that became an invaluable tool in research into the properties and effects of vacuums. He is remembered for the Magdeburg hemispheres: two half-spheres of metal, which, put together and the contained air evacuated, could not be separated even by two teams of eight horses.

Gutenberg, Beno (1889–1960) German-born American geophysicist whose research into seismology led him to propose a layer in the Earth's mantle in which seismic waves travel slowly. He measured the Earth's radius, concluding that its core was liquid.

Gutenberg, Johann (c. 1398–1468) German goldsmith noted for being the first to print a major work, the Gutenberg Bible, using separate letters (movable type) set in a frame for each page through the process of replica casting. It is often wrongly regarded as the dawn of printing. He developed special ink and transformed a wine press into a screw-and-lever press for printing pages.

Edmund Halley

Halley, Edmund (1656–1742) English astronomer, physicist, mathematician, seaman, and explorer who made many astronomical discoveries, but is best known for correctly predicting the return (in 1758, 1835, and 1910) of the comet that is now named after him. Halley was the first to propose that nebulas were clouds of interstellar gas in which stars were being formed. He was also the first to make a complete observation of the transit of Mercury.

Hamilton, Sir William Rowan (1805–65) Irish mathematician noted for his work in the field of complex numbers (numbers having a real part and an imaginary part containing the square root of –1). He proposed that some complex numbers could represent rotation in a plane. He also invented quaternions (generalized complex numbers with four components).

John Harrison

Harrison, John (1693–1776) English horologist and instrument maker who devised a chronometer of such accuracy that it could be used to determine longitude at sea to just over one minute of angle (1/60 of a degree, 18 geographical miles – 29 km). This clock won the prize of £20,000 that had been offered by the British Admiralty for an accurate chronometer. Harrison and his competitors for the prize became the subject of a best-selling book, *Longitude* (1995), by Dava Sobel. He also invented the gridiron pendulum.

Hauksbee, Francis (d. c. 1713) English physicist who is reputed to have made the first electrical machine and who determined the relative weights of air and water. His son, Francis Hauksbee (1687–1763), was a writer on scientific matters.

Hawking, Stephen William (b. 1942) British theoretical physicist and professor of mathematics whose work has greatly advanced our knowledge of space-time, black holes – singularities in space-time whose mass is so great that no light can escape – and the quantum theory of gravity. Hawking is best known for his best-selling book, *A Brief History of Time*, and for continuing to work despite suffering from Lou Gehrig's disease (motor neuron disease), which has produced almost complete paralysis.

Hermann Ludwig Ferdinand von Helmholtz

Heisenberg, Werner Karl (1901–76) German physicist who developed quantum mechanics and showed that it is impossible to know simultaneously with complete accuracy both the position and the momentum of a particle. This is the celebrated Heisenberg uncertainty principle. He also applied matrix calculus to atomic theory. Heisenberg was awarded the Nobel Prize in physics in 1932. His book *The Physical Principles of the Quantum Theory* (1949) is a classic.

von Helmholtz, Hermann Ludwig Ferdinand (1821–94) German physicist, physiologist, mathematician, and polymath who invented the ophthalmoscope, investigated the physics of music and optics, worked out the basis of color vision, and showed the equivalence of energy in food intake, muscular action, and heat production. The three-color (trichromatic) theory of color vision is known as the Young-Helmholtz theory.

Henry, Joseph (1797–1878) American physicist who, independent of Faraday, discovered electromagnetic induction and, in 1829, produced the first electric motor. He invented the large-scale electromagnet for industrial use and the electric relay. He showed the effect of electrical resistance in controlling electric current and showed that electric discharges are oscillatory in nature. The unit of inductance, the henry, is named after him.

Joseph Henry

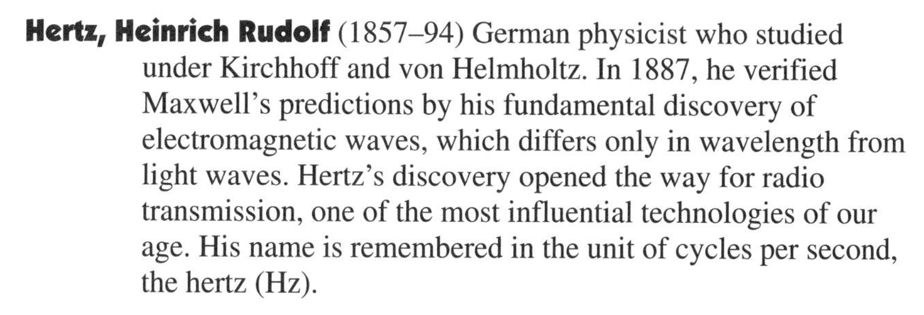

Hertz, Heinrich Rudolf (1857–94) German physicist who studied under Kirchhoff and von Helmholtz. In 1887, he verified Maxwell's predictions by his fundamental discovery of electromagnetic waves, which differs only in wavelength from light waves. Hertz's discovery opened the way for radio transmission, one of the most influential technologies of our age. His name is remembered in the unit of cycles per second, the hertz (Hz).

Hess, Harry H. (1906–69) American geologist and geophysicist who described the ocean floors and showed how they were constantly spreading as a result of the upthrust of magma into the mid-ocean ridges.

Harry H. Hess

Hess, Victor Francis (1883–1964) Austrian-born American physicist who compared the background radiation at the top of the Eiffel Tower in Paris with that at ground level and discovered that the former was higher. This suggested that some radiation was probably coming from outer space. Later investigations using balloons proved Hess was right. He had discovered cosmic rays. Hess shared the Nobel Prize in physics with Anderson in 1936.

Hofstadter, Robert (1915–90) American physicist who discovered that protons and neutrons were not elementary particles but had inner structures. He shared the 1961 Nobel Prize in physics.

Hollerith, Herman (1860–1929) American engineer who, while working on ways of handling the enormous task of the U.S. census, invented the punched card system of recording data and developed machines to punch the cards and then to tabulate, sort, analyze, and elicit information using these standard cards. The Hollerith Tabulating Machine Company thrived and Hollerith sold out for $1.25 million to a company that, in 1924, became IBM.

Holmes, Arthur (1890–1965) English geologist and geophysicist who put dates to the geological time scale. He determined the ages of rocks by measuring their radioactive constituents and was an early scientific supporter of Wegener's continental drift theory. He was the first to recognize that the Earth's crust formed (solidified) about 4.56 billion years ago. His book *Principles of Physical Geology* (1944) was highly successful.

Hooke, Robert (1635–1703) English physicist, chemist, and architect. One of the most brilliant scientists of his age and one of the most quarrelsome, he formulated Hooke's law of the extension and compression of elastic bodies and effectively invented the microscope and the first Gregorian telescope.

Hopper, Grace M. (1906–92) U.S. Navy Rear-Admiral and pioneer computer programer best remembered for her introduction of the term *bug* in a computer context. The original machines were electromechanical, using relays, and the bugs, which got between the contacts, were entomological. Hopper wrote the first software compiler in 1952 and had a major part in writing COBOL, the high-level language. She remained on active naval duty until she was age 80.

Hounsfield, Sir Godfrey Newbold (b. 1919) English engineer who invented the computer-assisted tomography scanner (CAT scanner). Tomography is a method of taking X-ray pictures in slices. Hounsfield showed how the results of many narrow-beam X-ray exposures at different angles within a slice could be analyzed by computer so as to determine the density at many points, and thus build up a picture.

Christiaan Huygens

Huygens, Christiaan (1629–93) Dutch physicist who first propounded the undulatory theory of light and discovered polarization. The "principle of Huygens" is part of wave theory. He also discovered the laws of the collision of elastic bodies simultaneously with John Wallis and Sir Christopher Wren.

Jansky, Karl G. (1905–50) American radio engineer who, while using a short-wave radio receiver with a directional antenna, noticed a background hiss that rose to a peak every 24 hours. He assumed, correctly, that this was due to the rotation of the Earth and that the signal was coming from outer space. He was then able to localize the source to the constellation of Sagittarius, near the center of the Milky Way. This was the dawn of radio astronomy. The unit of radio emission strength, the jansky, was named for him.

Jeffreys, Sir Harold (1891–1989) British geophysicist and astronomer whose treatment of mathematical physics applied to the Earth in his book *The Earth* (1924) was widely influential. He investigated the thermal history of the Earth, stresses in its crust, tidal friction, and the effect of the Earth's liquid core on nutation. His rejection of Wegener's theory of continental drift remains controversial.

Joliot-Curie, Jean Frédéric (1900–58) French physicist who, working with his wife, Irène Joliot-Curie (1897–1956), the daughter of Marie and Pierre Curie, laid the groundwork for the discovery of the neutron, and, by bombarding light elements with alpha particles (helium nuclei), produced the first artificial radioactive elements. For this work, they shared the Nobel Prize in chemistry in 1935.

Irène Joliot-Curie

Joule, James Prescott (1818–89) English physicist who laid foundations for the theory of the conservation of energy. He is famous for experiments in heat, which he showed to be a form of energy. He also showed that if a gas expands without performing work, its temperature falls. The joule, a unit of work or energy, is named after him. With Lord Kelvin, he devised an absolute scale of temperature.

James Prescott Joule

Kapitsa, Peter Leonidovich (1894–1984) Russian physicist who achieved unprecedentedly strong magnetic fields in his work on the deflection of alpha particles and magnetostriction. He also worked on the behavior of matter at low temperatures.

His work led to the discovery of superfluidity (zero viscosity in one direction). He shared the Nobel Prize in physics for 1978.

Karle, Jerome (b. 1918) American physicist who, with Herbert Hauptman, developed a new method of the rapid interpretation of X-ray crystallography data, work for which they were awarded the 1985 Nobel Prize in chemistry.

von Karman, Theodore (1881–1963) Hungarian-born American physicist who was responsible for the setting up of the Jet Propulsion Laboratory at the California Institute of Technology. He was a world authority on aerodynamics and described the two rows of vortices in the wake behind a body moving in a fluid, a phenomenon known as the Karman vortex street.

William Thomson, Lord Kelvin

Kelvin of Largs, William Thomson, 1st Baron (1824–1907) Irish-born Scottish physicist and mathematician who proposed the absolute, or Kelvin, temperature scale (1848) and, at the same time as Clausius, established the second law of thermodynamics. He also invented a tide predictor and a harmonic analyzer.

Kerr, John (1824–1907) Scottish physicist who discovered that if a beam of plane-polarized light is reflected from the polished end of an energized electromagnet, its axis will be rotated and it will become elliptically polarized. This is called the magneto-optical, or Kerr, effect.

Kerr, Roy Patrick (b. 1934) New Zealand mathematician and astrophysicist who solved Albert Einstein's field equations of general relativity and provided support for, and a complete description of, rotating black holes.

Gustav Robert Kirchoff

Kirchhoff, Gustav Robert (1824–87) German physicist who, while still a student, devised the laws, now known as Kirchhoff's laws, for determining currents in electrical networks. Working in spectroscopy, he discovered the elements cesium and rubidium and identified elements in the solar atmosphere. He formulated Kirchhoff's laws of radiation, which stipulate that, for a given wavelength, the ratio of emission to absorption is the same for all bodies at a given temperature.

von Kleist, Ewald Georg (c. 1700–48) Prussian (German) physicist who invented the Leyden jar independently of others. The Leyden jar was the first electric capacitor, formerly called "condenser," in which the glass of the jar formed the dielectric and metal foil on the inside and outside formed the plates. The original jar used water as the inner conductor. Leyden jars could store a large electric charge.

Lagrange, Joseph Louis (1736–1813) French mathematician noted for his skill in finding general solutions for equations and problems. He contributed significantly to number theory, differential equations, calculus of variations, and mechanics. A Lagrangian function is one used to define a dynamic system in terms of functions of coordinates, velocities, and times.

Joseph Louis Lagrange

Land, Edwin Herbert (1909–91) American inventor who developed the useful eyeshade plastic, Polaroid. This consisted of millions of tiny polarizing crystals in a plastic matrix. In 1947, Land also invented a new kind of color camera, which he called the Polaroid-Land camera.

Landau, Lev Davidovich (1908–68) Azerbaijan (Soviet) physicist noted for his research and publications in low-temperature physics, solid-state physics, nuclear physics, fluid dynamics, thermodynamics, quantum field theory, and astrophysics. He discovered superfluidity (zero viscosity in one direction) in liquid helium and superconductivity in the same element, for which he was awarded the Nobel Prize in physics in 1962.

Lev Davidovich Landau

Landé, Alfred (1888–1975) German physicist who in 1922 demonstrated the splitting of the D lines of sodium on the spectroscope into four or more components by a strong magnetic field. Studies of this kind led to the exclusion principle.

Langevin, Paul (1872–1946) French physicist who investigated the nature of magnetism and showed that paramagnetism (the weak susceptibility to be attracted by a magnetic field) could be explained in terms of orbital electrons. He also showed that paramagnetism was a function of absolute temperature. Langevin invented sonar, using piezoelectric transducers.

Irving Langmuir

Langmuir, Irving (1881–1957) American physicist and inventor who greatly improved the performance and longevity of tungsten electric lamps by developing a coiled filament and filling the bulb with inert gas. His most important scientific work was in connection with adsorption onto surfaces. This advanced the understanding of catalysts. He also did research on proteins, welding, gaseous discharges, and oil films on water. For his work on surface chemistry he was awarded the Nobel Prize in chemistry in 1932.

Laplace, Pierre Simon (1749–1827) French mathematician, astronomer, and physicist who proposed a theory of the origin of the solar system and developed celestial mechanics. He also worked on probability, Moon and planetary perturbations, and the tides. He showed that planetary perturbations caused by the gravitational effect of other planets would not disturb their orbits. He is best remembered for Laplace transforms, a method of converting one function into another function of a different variable.

Larmor, Sir Joseph (1857–1942) Irish physicist who criticized Maxwell's theories of the nature of electromagnetism and was one of the first to develop a theory explaining the magnetic fields of the Earth and the Sun.

Ernest Orlando Lawrence

Lawrence, Ernest Orlando (1901–58) American physicist who invented the cyclotron, a machine that achieved high particle acceleration in a comparatively compact space by using a spiral path. This enabled him to transmute elements and to produce hundreds of new radioactive isotopes. For this work, Lawrence was awarded the Nobel Prize in physics in 1939.

Lederman, Leon M. (b. 1922) American physicist who confirmed that the muon was not simply a heavy electron but a combination of two neutrinos. He also discovered the neutral kaon and the "bottom" quark. Lederman shared the Nobel Prize in physics in 1988.

Lee, David (b. 1931) American physicist who, working with Robert Richardson and Osheroff, discovered superfluidity in helium-3 at very low temperatures. This advance earned the three men the 1996 Nobel Prize in physics.

Lippershey, Hans (c. 1570–c. 1619) Dutch optician and spectacle maker who showed that a combination of a short-focus convex eyepiece lens and a long-focus convex objective lens, separated by an appropriate distance, could produce an instrument that made distant objects appear to be closer. Lippershey applied for a patent for this device in 1608 but was probably not the only investigator to make the discovery of the telescope.

London, Heinz (1907–70) German-born British physicist who studied superconductivity and evolved equations for the electromagnetic properties of superconductors. He achieved a temperature of within one-tenth of a degree above absolute zero.

Lorentz, Hendrik (1853–1928) Dutch physicist who, with George Fitzgerald, in 1895 proposed a mathematical transformation, known as the Fitzgerald-Lorentz contraction, to account for the failure of the Michelson-Morley experiment to show relative motion between the moving earth and the postulated ether. The contraction is that of any moving body in the direction of its motion and is negligible unless the speed approaches that of light. This idea helped to lead to Einstein's special theory of relativity. He shared the 1902 Nobel Prize in physics with Zeeman.

Hendrik Lorentz

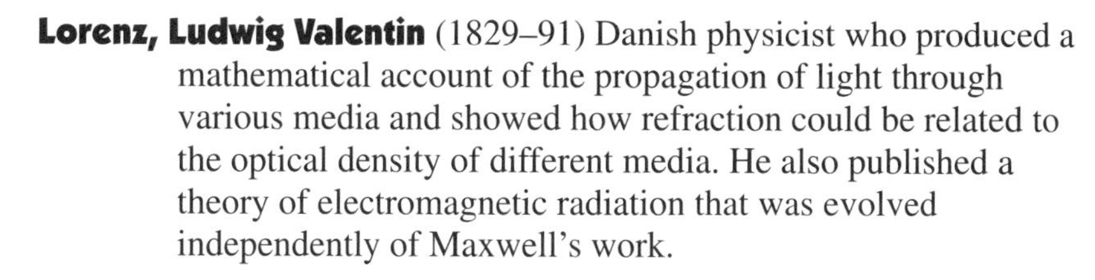

Lorenz, Ludwig Valentin (1829–91) Danish physicist who produced a mathematical account of the propagation of light through various media and showed how refraction could be related to the optical density of different media. He also published a theory of electromagnetic radiation that was evolved independently of Maxwell's work.

Love, Augustus (1863–1940) English geophysicist who was the first to detect seismic waves transmitted over the surface of the Earth. These are called "Love waves." His work contributed notably to the assessment of the differences in thickness of the Earth's crust below the continents and below the oceans.

Lovelace, Ada, Countess of (1815–52) English mathematician and daughter of the poet Byron who, after translating from the Italian an account of Babbage's mechanical analytical engine, became the first computer buff. She clearly understood what Babbage was aiming to do and added some valuable suggestions of her own. She is remembered in the high-level programing language called ADA.

Ernst Mach

Mach, Ernst (1838–1916) Austrian physicist and philosopher whose writings influenced Einstein. He contributed to the science of projectiles and gave names to the ratio of the speed of flow of a gas to the speed of sound (Mach number) and to the angle of a shock wave to the direction of motion (Mach angle).

McMillan, Edwin M. (1907–91) American nuclear physicist who, working with Philip Abelson, bombarded uranium with neutrons, and produced the first human-made element, atomic number 93, which they called neptunium. The two men were awarded the 1951 Nobel Prize in chemistry.

Magnus, Heinrich Gustav (1802–70) German physicist and chemist who studied tellurium and selenium and the gases in the blood. He is best remembered for his discovery of the fact that a flow of air over a rotating cylinder produces a sideways force. This effect, the Magnus effect, has been used to propel sailing boats and is well known to high-handicap golfers.

Theodor Maiman

Maiman, Theodore (b. 1927) American physicist who was the first to make a functioning laser. Coherent microwave generation had been achieved, by American physicist Charles Townes, and the principle established, but Maiman extended this work into the optical region, thereby producing a device with a very large range of practical and research applications. The great bulk of recorded music and much recorded digital information is now reproduced using a laser, and there are many communications, engineering, commercial, and surgical applications.

Marconi, Guglielmo, Marchese (1874–1937) Italian inventor who, having learned of Hertz's production of long wavelength electromagnetic waves (below the infrared), began to experiment in their detection and their use for sending Morse code messages. Marconi's transmitter was an induction coil producing a high-voltage, high-frequency alternating current across a spark gap, and the detector was a "coherer" (a crude rectifier). A ground connection was made at both transmitter and receiver, and long wire antennas were used. Marconi's persistence as an experimenter in the face of official indifference eventually showed the practicality of long-distance radio communication. He shared the Nobel Prize in physics in 1909.

Maxwell, James Clerk (1831–79) Scottish physicist, and one of the greatest theoretical physicists, whose work on the theory of electromagnetic radiation is considered to have paved the way for Einstein and Planck. He provided the mathematical treatment of his friend Faraday's theory of electrical and magnetic forces, showed that light was an electromagnetic wave phenomenon, and produced mathematical expressions for scientific findings in optics, the kinetic theory of gases, and electromagnetism. His work transformed physical science.

Meitner, Lise (1878–1968) Austrian nuclear physicist who discovered the radioactive element protoactinium and made the telling and correct suggestion that the presence of radioactive barium in the products of uranium that had been bombarded with neutrons was probably due to the fact that some uranium atom nuclei had been split in two. The transuranic element number 108, meitnerium, has been named for her.

Lise Meitner

Michelson, Albert Abraham (1852–1931) Polish-born American physicist and the first American scientist to win a Nobel Prize, awarded in 1907, for determining the speed of light. He is chiefly remembered, however, for the Michelson-Morley experiment using an optical interferometer, which he and Morley had invented and which, although adequately sensitive, failed to show the existence of the ether. This result led to the abandonment of the mistaken concept of space being filled with ether.

Millikan, Robert Andrews (1868–1953) American physicist noted for measuring *e* (the charge on the electron) and then *E* (the energy of an electron). He also studied, and named, cosmic rays. He was awarded the Nobel Prize in physics in 1923.

Robert Andrews Millikan

Morley, Edward William (1838–1923) American chemist and physicist who, with Michelson, developed a sensitive interferometer with which they showed that the speed of light is constant whether measured in the direction of the Earth's movement or perpendicular to that direction (Michelson-Morley experiment).

Moseley, Henry Gwyn Jeffreys (1887–1915) British physicist who used X-ray scattering by different elements to show that the resulting wavelengths decreased regularly with the increase of atomic weight. He concluded, correctly, that each element had a different number of electrons.

John Napier

Mueller, Erwin Wilhelm (1911–77) German-born American physicist who invented the field-emission needle-tip microscope capable of magnifications of more than a million times. Under it the positions of individual atoms in a crystal structure can be seen.

Napier, John (1550–1617) Scottish mathematician who invented logarithms and produced an early form of slide rule known as "Napier's bones." He spent 20 years writing tables of logarithms, unfortunately not to the base 10, and published his logarithm method of computation in a book in 1614. Napier also calculated the orbits of planets.

Newton, Sir Issac (1642–1727) English mathematician and scientist. By 1684 he had demonstrated his famous gravitation theory. Independently of Leibnitz, he discovered the differential calculus. He also discovered that white light is composed of many colors and invented a reflecting telescope. His great work, *Philosophiae naturalis principia mathematica* (1687), established him as the leading scientist of his day.

Nicol, William (c. 1768–1851) Scottish physicist who glued two crystals of Iceland spar together to produce the "Nicol prism," which produced plane-polarized light. Two Nicol prisms at right angles to each other would cut off the light altogether. This device originated the discipline of polarimetry and was of wide application in physics.

Nollet, Jean (1700–70) French physicist who as a student gave up theology to devote his life to science. He worked mainly in electricity and invented an electroscope and an improved Leyden jar (capacitor). He also clearly explained osmosis.

Oersted, Hans Christian (1777–1851) Danish physicist who showed that the passage of an electric current along a conductor was always associated with a magnetic field around the conductor. He demonstrated this by showing the deflection of a compass needle near a wire whenever a current flowed through the wire. The discovery prompted much research into electricity.

Hans Christian Oersted

Ohm, Georg Simon (1789–1854) German physicist whose studies of how the current flowing in an electric circuit was affected by the applied voltage and the resistance led to the formulation of the law that has ever since honored his name. Because there are three variables, the law can be stated in three ways. He is also remembered in the name of the unit of resistance, the ohm.

Georg Simon Ohm

Oppenheimer, J. Robert (1904–67) American physicist whose studies into quantum mechanics helped to provide theoretical physics with a mathematical basis. He predicted the existence of a positive anti-electron, which was later discovered, and he proved that black holes would form when the thermonuclear energy of massive stars was exhausted and gravitational forces could start the compression process. He is, however, best known for his major contribution, as director of the Los Alamos laboratories, to the design of the atomic bomb.

Osheroff, Douglas (b. 1945) American physicist who, working with Robert Richardson and David Lee, discovered superfluidity in helium-3 at very low temperatures. This advance earned the three men the 1996 Nobel Prize in physics.

Oughtred, William (c. 1574–1660) English mathematician who introduced the sign for multiplication and the trigonometrical abbreviations "sin," "cos," "tan" for "sine," "cosine," and "tangent." He, as well as Napier, was an inventor of the slide rule, which he produced by laying two logarithmic scales side by side.

Pascal, Blaise (1623–62) French mathematician and physicist who originated the mathematics of probability, invented a mechanical adding and subtracting machine, developed projective geometry from work on conic sections, and proved that atmospheric pressure declined with altitude. Pascal invented a mechanical calculator, and his name is remembered in the computer programming language Pascal.

Paschen, Louis (1865–1947) German physicist who proved that the helium found on Earth is the same as that on the Sun. He is remembered by a series of lines in the hydrogen spectrum that were named after him.

Wolfgang Pauli

Pauli, Wolfgang (1900–58) Austrian-born American theoretical physicist who showed that no two electrons in an atom can be in the same quantum state – a fact that goes far to account for the electron structure of the atom. This is known as the Pauli exclusion principle, and it earned Pauli the Nobel Prize in physics in 1945. In 1957 Pauli carried out experiments confirming the nonparity theory of Yang and Tsung Dao Lee in nuclear interaction.

Peltier, Jean Charles Athanase (1785–1845) French physicist who discovered in 1834 that the junction of two dissimilar metals (such as bismuth and antimony) in an electric circuit became hotter or colder depending on the direction of flow of the electric current. This became known as the Peltier effect.

Perey, Margeurite Catherine (1909–75) French physicist who worked under Marie Curie, and later, in 1939, discovered the element francium. She was the first woman to become a member of the French Academy of Sciences. Like her former chief, she died from a malignant disease that was almost certainly induced by overexposure to radiation.

Perl, Martin L. (b. 1927) American physicist who, with Reines, discovered the neutrino and the tau lepton. The demonstration of the existence of the neutrino has been described as "a feat bordering on the impossible," but these two researchers succeeded, thereby greatly advancing particle physics. For this work, they were awarded the 1995 Nobel Prize in physics.

Phillips, William D. (b. 1948) American physicist who, working independently of Chu, developed a method of using laser light to cool and trap atoms, allowing them to be studied in detail. For this work, which has greatly extended knowledge of the relationship of matter and radiation and which promises important practical applications, Phillips and Chu were awarded the Nobel Prize in physics in 1997, shared with Cohen-Tannoudji.

Planck, Max Karl Ernst Ludwig (1858–1947) German theoretical physicist who in 1900, in an attempt to provide a theoretical basis for the "ultraviolet catastrophe" paradox in black-body radiation, proposed the seemingly ridiculous idea that radiation was emitted and received in packets (quanta). In this way, quantum theory was born. His name is remembered in the term *Planck's constant*, for a very small number, h, in the equation $E = hv$, which relates the energy, E, of a quantum to its frequency, v. Planck was awarded the Nobel Prize in physics in 1918.

Poincaré, Jules Henri (1854–1912) French mathematician who contributed notably to many branches of mathematics and who, independently of Einstein, came close to establishing a relativity theory. He worked on topology, theory of numbers, probability theory, automorphic functions, and celestial mechanics, but even so, was liable to make mistakes in elementary arithmetic.

Siméon Denis Poisson

Poisson, Siméon Denis (1781–1840) French mathematician and mathematical physicist who advanced ideas in the transformation of equations in mechanics, defined integrals, and mathematically described electromagnetic theory and probability. He produced a special case of the binomial distribution in statistics, known as the Poisson distribution.

Powell, Cecil Frank (1903–69) English physicist who improved on the Wilson cloud chamber as a means of displaying particle tracts by developing a special photographic emulsion that would do the job better and more easily. A plate developed from this idea was exposed in a high-altitude balloon to investigate cosmic rays and revealed a new meson particle, which was named the pi-meson, or pion. This particle had been predicted by Yukawa, and its discovery won Powell the 1950 Nobel Prize in physics.

Ludwig Prandtl

Prandtl, Ludwig (1875–1953) German physicist who studied fluid mechanics and discovered that a liquid flowing in a tube has a thin boundary layer that moves more slowly than the rest of the liquid. This discovery of skin friction led to advances in streamlining. In 1909 Prandtl helped to build the first German wind tunnel.

Purcell, Edward Mills (1912–97) American physicist who worked in radar, radio astronomy, astrophysics, and biophysics but whose main contribution was in nuclear spin resonance and, with Bloch, in the development of nuclear magnetic resonance scanning. This has become important in science, medicine, and engineering. Purcell and Bloch were awarded the 1952 Nobel Prize in physics.

Pythagoras (c. 582–c. 497 BCE) Greek philosopher and physicist who made notable advances in the physics of sound, especially the sound produced by the plucking of stretched strings. He showed that the square root of two could not be expressed as the ratio of two numbers (a whole fraction) and proved that the area of the square on the hypotenuse of a right-angled triangle is equal to the sum of the areas of the squares on the other two sides.

Rabi, Isidor Isaac (1898–1988) Austro-Hungarian-born American physicist who developed the magnetic resonance method of observing spectra. This advance allowed confirmation of the theory of quantum electrodynamics and led to the development of medical and industrial scanners of higher resolution than CT scanners. For this work, Rabi was awarded the Nobel Prize in physics in 1944.

Raman, Sir Chandrasekhar Venkata (1888–1970) Indian physicist who showed that light scattered by the molecules of a liquid acquired higher and lower frequencies (the Raman effect) from the absorption or donation of energy by the molecular bonds. This work was one of the first independent confirmations of quantum theory and led to an improved method of analyzing molecular structure, known as Raman spectroscopy. For this work, he was awarded the Nobel Prize in physics in 1930.

Rankine, William John M. (1820–72) Scottish physicist and engineer who worked on heat and thermodynamics, coined the terms *potential energy* and *actual energy* (kinetic energy), and, in his popular textbooks, successfully bridged the gap between science and engineering.

Rayleigh, John William Strutt, 3rd Baron (1842–1919) English physicist and mathematician who wrote a classic work on acoustics, *The Theory of Sound* (1879). He worked on optics, standardized the ohm and the amp, produced a formula for black-body radiation (that failed to account for high frequencies because quantum theory had not yet emerged), and helped William Ramsay to discover the element argon. In 1904 he was awarded the Nobel Prize in physics and Ramsay the Prize in chemistry.

John William Strutt, Lord Rayleigh

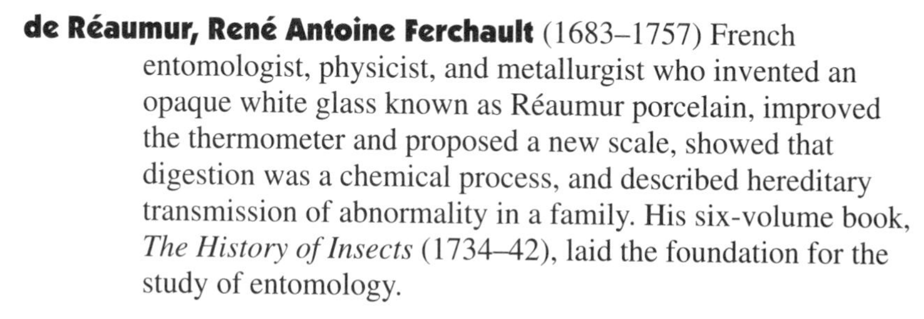

de Réaumur, René Antoine Ferchault (1683–1757) French entomologist, physicist, and metallurgist who invented an opaque white glass known as Réaumur porcelain, improved the thermometer and proposed a new scale, showed that digestion was a chemical process, and described hereditary transmission of abnormality in a family. His six-volume book, *The History of Insects* (1734–42), laid the foundation for the study of entomology.

René Antoine Ferchault de Réaumur

Reich, Ferdinand (1799–1882) German physicist who, working with Hieronymous Richter, discovered the soft, silvery metal indium. He used spectroscopy to find the new element. Indium is used in fusible alloys and in some semiconductor applications.

Reines, Frederick (b. 1918) American physicist who, with Perl, discovered the neutrino and the tau lepton. The neutrino, in particular, was extremely difficult to demonstrate because of the paucity of its characteristics – the absence of charge and the zero rest mass – but Reines and Perl succeeded. For this work, they were awarded the 1995 Nobel Prize in physics.

Richardson, Sir Owen Willans (1879–1959) British physicist who explained why heated metals emit electrons and that the rate of emission was proportional to the absolute (Kelvin) temperature of the metals. This was the Edison effect, first discovered by Thomas Alva Edison, and it proved to be of great importance, as it permitted the development of the thermionic valve, early radio, and the cathode-ray tube, television, and personal

Sir Owen Willans Richardson

computers. For his work on thermionics, Richardson was awarded the 1928 Nobel Prize in physics.

Richardson, Robert (b. 1937) American physicist who, working with David Lee and Osheroff, discovered superfluidity in helium-3 at very low temperatures. This advance earned the three men the 1996 Nobel Prize in physics.

Georg Friedrich Bernard Riemann

Riemann, Georg Friedrich Bernard (1826–66) German mathematician who, as a young man, described a geometry of an n-dimensional curved space in a nervous lecture unexpectedly demanded by the celebrated and intimidating Gauss. He greatly extended non-Euclidean geometry and laid a foundation for Einstein's special theory of relativity.

Ritter, Johann (1776–1810) German physicist who discovered ultraviolet light. He performed electrolysis of water, using a galvanic battery, and collected the resulting gases. He invented the first electric dry cell and the first chargeable battery, and understood electricity in terms of chemical action.

Wilhelm Conrad Roentgen

Roentgen, Wilhelm Conrad (1854–1923) German physicist who noticed that when he turned on a gas discharge tube, even in the dark, a card coated with a fluorescing compound glowed. He had discovered X rays. He quickly demonstrated their medical value, but not their dangers. In 1901, he was awarded the first Nobel Prize in physics.

Ruhmkorff, Heinrich (1803–77) German-born French instrument maker who invented an induction coil consisting of a core of soft iron around which was wound a primary coil of a small number of turns of thick wire and a secondary winding of a large number of turns of thinner wire. Current in the primary winding was interrupted by an electromechanical interrupter mechanism, and the result was a very high voltage in the secondary winding. This coil was used for early radio transmission, X-ray tubes, cathode-ray tubes, etc.

Rumford, Benjamin Thompson, Count (1753–1814) American-born British adventurer and physicist who, while watching the boring of cannon for his benefactor, the elector of Bavaria, noticed that heat continued to be produced as long as the boring continued. He concluded that the current caloric theory of heat was wrong and that heat was a kind of motion.

Rutherford, Ernest (1st Baron Rutherford of Nelson) (1871–1937) New Zealand–born British pioneer of subatomic particle physics whose work led to the revolutionary concept of the atom as a miniature solar system, with a nucleus surrounded by planetary electrons. He correctly predicted the existence of the neutron.

Rydberg, Johannes Robert (1854–1919) Swedish physicist who derived a formula for the quantitative relationships of the frequencies of the spectroscopic lines. This involved a constant that was later confirmed by quantum studies of the atom. He also proposed that the elements should be listed by atomic number, not by atomic weight.

Sakharov, Andrei Dmitriyevich (1921–89) Russian nuclear physicist and particle physicist who was the principal figure in the development of the Soviet hydrogen bomb, first tested in 1962. Sakharov was heaped with honors but soon became a human-rights political activist and critic of Stalin's regime. Although awarded the Nobel Peace Prize, he was arrested and exiled to Gorky in 1980. He was freed in 1986 by Gorbachev.

Andrei Dmitriyevich Sakharov

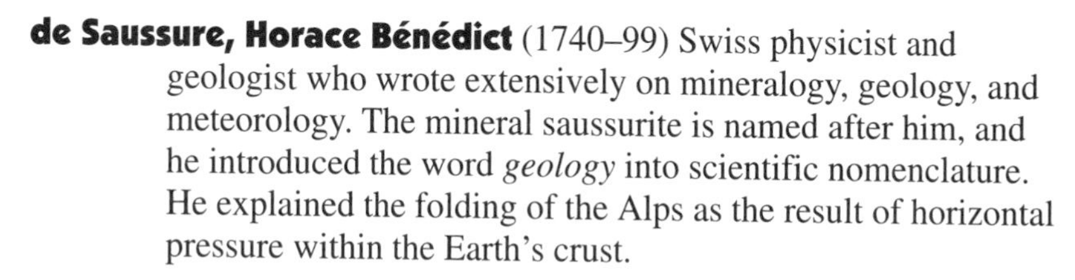

de Saussure, Horace Bénédict (1740–99) Swiss physicist and geologist who wrote extensively on mineralogy, geology, and meteorology. The mineral saussurite is named after him, and he introduced the word *geology* into scientific nomenclature. He explained the folding of the Alps as the result of horizontal pressure within the Earth's crust.

Horace Bénédict de Saussure

Schrödinger, Erwin (1887–1961) Austrian physicist whose early papers originated the science of quantum wave mechanics. After de Broglie suggested the identity of particles and waves, Schrödinger evolved a wave equation describing the behavior of such systems. Schrödinger also described the quantum behavior of electrons. He shared with Dirac the 1933 Nobel Prize in physics.

Schweigger, Johann (1779–1857) German physicist who invented an improved and more sensitive galvanometer, which influenced others to improve it even further.

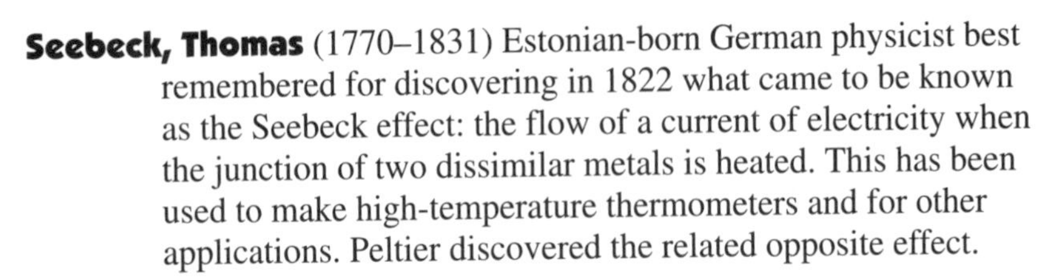

Seebeck, Thomas (1770–1831) Estonian-born German physicist best remembered for discovering in 1822 what came to be known as the Seebeck effect: the flow of a current of electricity when the junction of two dissimilar metals is heated. This has been used to make high-temperature thermometers and for other applications. Peltier discovered the related opposite effect.

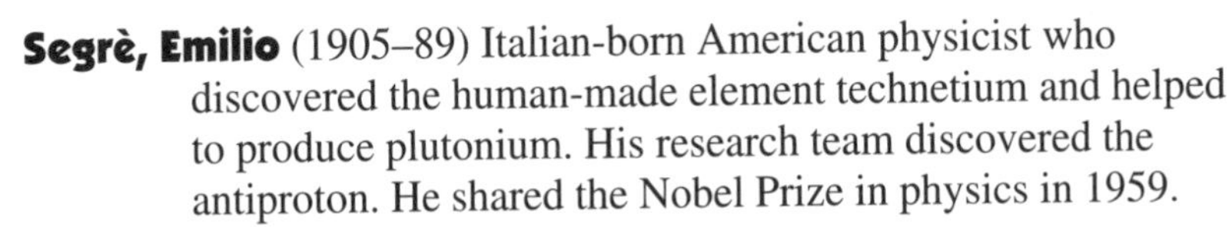

Segrè, Emilio (1905–89) Italian-born American physicist who discovered the human-made element technetium and helped to produce plutonium. His research team discovered the antiproton. He shared the Nobel Prize in physics in 1959.

Nikolai Nikolaevich Semenov

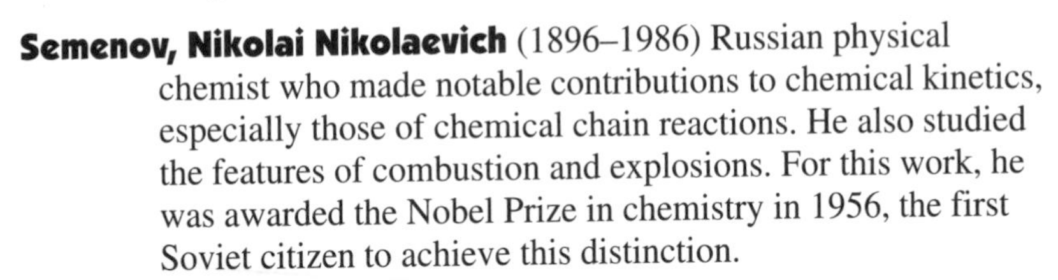

Semenov, Nikolai Nikolaevich (1896–1986) Russian physical chemist who made notable contributions to chemical kinetics, especially those of chemical chain reactions. He also studied the features of combustion and explosions. For this work, he was awarded the Nobel Prize in chemistry in 1956, the first Soviet citizen to achieve this distinction.

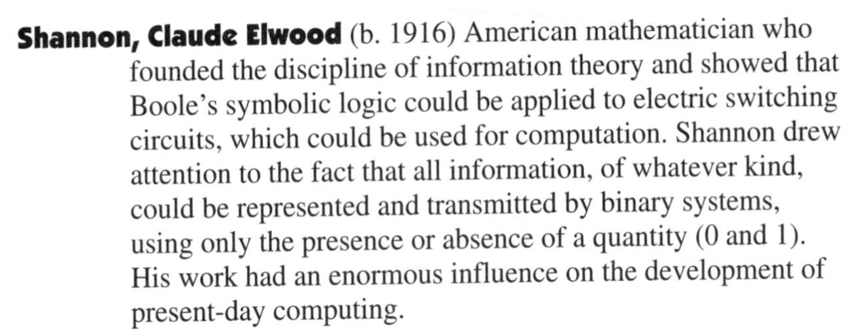

Shannon, Claude Elwood (b. 1916) American mathematician who founded the discipline of information theory and showed that Boole's symbolic logic could be applied to electric switching circuits, which could be used for computation. Shannon drew attention to the fact that all information, of whatever kind, could be represented and transmitted by binary systems, using only the presence or absence of a quantity (0 and 1). His work had an enormous influence on the development of present-day computing.

Shockley, William (1910–89) American physicist who worked at Bell Laboratories with Bardeen and Brattain on solid-state substitutes for thermionic valves. They produced a germanium point-contact rectifier and then added another electrode to produce the first transistor amplifier. Junction transistors quickly followed, and the world was never the same again. The three men shared the 1956 Nobel Prize in physics.

Shull, Clifford G. (b. 1915) American physicist who, working independently of Brockhouse, developed the powerful technique of neutron scattering. This uses beams of neutrons, which, when impinging on the nuclei of the atoms of matter, diffract and produce a pattern that provides much detailed information. The method allows the investigation of the atomic structure of bulk matter and is a technique of high theoretical and practical importance. It earned Shull and Brockhouse the 1994 Nobel Prize in physics.

Snell, Willebrod (1580–1626) Dutch mathematician and physicist who discovered how the angles of incidence and refraction of a ray of light passing from a medium of one density to one of another density were related and were determined by the refractive indices of the two media. This is called Snell's law and equates the sines of the angles multiplied by the refractive indices. He tried to use triangulation to survey the earth.

Willebrod Snell

Stark, Johannes (1874–1957) German physicist who discovered the splitting of spectral lines by a strong magnetic field and described the Doppler effect operating in rays of charged particles emitted by the anodes of discharge tubes. He was awarded the 1919 Nobel Prize in physics. Stark was an aggressive, anti-Semitic Nazi who supported Hitler and turned against quantum theory and relativity, which he described as "Jewish science."

Stefan, Josef (1835–93) Austrian physicist who deduced the law that the radiant energy emitted each second from a black body is proportional to the fourth power of the absolute temperature. This is known as Stefan's law. He also calculated the temperature of the Sun's surface.

Stern, Otto (1888–1969) American physicist who proved a theory of quantum physics experimentally. He projected a beam of silver atoms through a nonuniform magnetic field and showed that it separated into two distinct beams, indicating that an atom has a magnetic moment that can only be oriented in two definite directions relative to an external magnetic field. This proof won Stern the Nobel Prize in physics in 1943.

Stevinus, Simon (1545–1620) Flemish mathematician and military engineer who attacked the idea of perpetual motion, showed that hydrostatic pressure depended on depth and not on the shape of the vessel, and introduced the system of decimal fractions.

Simon Stevinus

Sir George Stokes

Stokes, Sir George (1819–1903) British mathematician and physicist who devised the mathematical description of a small spherical body falling through a viscous fluid. This was an early account of fluid flow viscosity. His equation (Stokes' law) has been found to have numerous practical and theoretical applications. Late in life, Stokes suggested correctly that X rays were probably electromagnetic in nature.

Sturgeon, William (1783–1850) British physicist who about 1823 invented the electromagnet by insulating a metal core with varnish and wrapping a wire around it. This advance antedated Faraday's great work on electromagnetic induction, but only just.

Swan, Sir Joseph Wilson (1828–1914) English chemist and physicist who in 1860 invented the electric lamp 20 years before Thomas Edison. In 1864 he patented the carbon process for photographic printing; in 1871 he invented the dry-plate technique; and in 1879 he produced bromide paper. He was the first to produce a practicable artificial silk.

Nikola Tesla

Tesla, Nikola (1856–1943) Croatian-born American electrical engineer who invented the alternating current induction motor, the polyphase AC system, and a high-voltage, air-cored, high-frequency transformer known as the Tesla coil, one version of which produced a 40-meter spark. Tesla was an eccentric recluse who was offered the Nobel Prize in 1912, but refused to accept it.

Thales (c. 624–545 BCE) Greek astronomer, mathematician, and physicist who is said to have correctly predicted a solar eclipse and who believed that everything was compounded from water.

Sir George Paget Thomson

Thomson, Sir George Paget (1882–1975) English physicist who was the first to notice that a beam of electrons passing though a thin foil of metal produced interference patterns. This was an important finding that confirmed de Broglie's theory that moving particles have wave properties. It also enabled Thomson, and Davisson in the United States, to look more deeply into the diffraction characteristics of electrons. This work earned the two the 1937 Nobel Prize in physics.

Thomson, Sir Joseph John (1856–1940) English pioneer of nuclear physics who discovered the electron. He demonstrated that "corpuscles" (electrons) comprising cathode rays were nearly 2,000 times smaller in mass than the then lightest-known particle, the hydrogen ion. This called for a revision of the model of the atom. Thomson imagined it as a positively charged mass, the nucleus, with electrons distributed around it, like planets around the Sun. In 1906, he was awarded the Nobel Prize in physics.

Torricelli, Evangelista (1608–47) Italian physicist and mathematician who was the first to discover the fundamental principles of hydraulics (fluid mechanics). In 1644 he described the barometer, or Torricellian tube. He improved telescopes and microscopes and made several mathematical discoveries. He is said to be the father of hydrodynamics.

Evangelista Torricelli

Townes, Charles Hard (b. 1915) American physicist who, in an attempt to generate electromagnetic radar waves of ever shorter wavelength (higher frequency), developed an oscillating device, using ammonia molecules, that produced coherent radiation. This was the maser (*m*icrowave *a*mplification by *s*timulated *e*mission of *r*adiation), and it won him a share of the 1964 Nobel Prize in physics with Basov and Aleksandr Prokhorov. It also led the way to the development by others of a similar device working at visible and near-visible light frequencies: the laser.

Tyndall, John (1820–93) Irish physicist best known for his work on the scattering of light by colloidal particles in solution (the Tyndall effect). He also worked on the transmission of infrared radiation through gases and vapors.

Van Allen, James Alfred (b. 1914) American physicist who developed the radio proximity fuse, employed rockets to study the physics of the upper atmosphere, contributed to the success of the first American artificial satellite, and discovered two belts of charged particles circling the earth (Van Allen belts), which are retained by the Earth's magnetic field.

van de Graaff, Robert Jemison (1901–67) American physicist who developed a very high-voltage static electricity generator that proved to be a powerful tool as a particle accelerator in atomic and nuclear physics.

Alessandro Giuseppe Anastasia Volta

Volta, Alessandro Giuseppe Anastasia, Count (1745–1827) Italian physicist and inventor of the first electric battery (the voltaic pile) in 1800 whose name is given to the unit of electrical potential difference, the volt. Volta developed a theory of current electricity and made studies of heat and gases.

van der Waals, Johannes Diderik (1837–1923) Dutch physicist who, aware that real gases did not accurately conform to the simple gas law $pV = RT$ (pressure x volume = temperature x the gas constant R), devised a more precise gas law equation that took account of the volume of the gas molecules and the attraction between them. He received the Nobel Prize in physics in 1910.

Sir Robert Watson-Watt

Walton, Ernest (1903–95) Irish physicist who, working with Cockcroft, was the first to produce the artificial fission of an atomic nucleus. He bombarded lithium with high-energy protons, and, by determining the energy of the alpha particles produced, was able to confirm Einstein's equation $E = mc^2$, showing the equivalence of mass and energy. Walton and Cockcroft shared the 1951 Nobel Prize in physics.

Watson-Watt, Sir Robert (1892–1973) Scottish physicist who used the reflection of short-wave radio waves from thunderclouds for meteorological purposes and then, by 1935, had applied the same idea to finding the location of flying aircraft. He called this "radio-location," and the system was hastily improved so that by the beginning of World War II, radio-location (radar) units were set up in various parts of Britain.

Sir Charles Wheatstone

Weiss, Pierre Ernst (1865–1940) French physicist who developed the theory of ferromagnetism, which held that iron and other ferromagnetic materials form domains of polarity that, once oriented in alignment, create a strong magnetic force.

Wheatstone, Sir Charles (1802–75) British physicist and inventor who created the first commercial electric telegraph. He is best known among engineers for his ingenious four-arm electrical bridge, the Wheatstone bridge, that can be used for the accurate measurement of electrical resistance and modified for the measurement of inductance.

Wigner, Eugene Paul (1902–95) Hungarian-born U.S. engineer and atomic physicist who emigrated to the United States and, in 1933, established facts on neutron capture that made it clear that an atomic bomb was feasible. This led to the decision of the United States to build the bomb. Wigner shared the Nobel Prize in physics in 1963.

Wilkins, Maurice Hugh Frederick (b. 1916) New Zealand-born British biophysicist who, with Crick, Watson, and Rosalind Franklin, worked to determine the molecular structure of DNA by X-ray crystallography. Wilkins shared the Nobel Prize in physiology or medicine with Crick and Watson in 1962.

Wilson, Charles Thomson Rees (1869–1959) Scottish pioneer of atomic and nuclear physics and researcher of atmospheric electricity who devised the cloud-chamber method of marking the track of alpha particles and electrons, thus allowing movement and interaction of atomic and subatomic particles to be followed and photographed. He shared the 1927 Nobel Prize in physics with Compton.

Charles Thomson Rees Wilson

Wilson, Kenneth G. (b. 1936) American theoretical physicist who is noted for his work on the application of ingenious mathematical techniques to the magnetic properties of atoms and to the changes of matter from the solid to the liquid and gas phases and the changes in ferromagnetism at the Curie point. His ideas were formulated into a general theory that won him the Nobel Prize in physics in 1982.

Yalow, Rosalyn Sussman (b. 1921) U.S. nuclear physicist who developed the technique of radioimmunoassay, which is a means of detecting and precisely measuring extremely small amounts of almost any substance to which antibodies are formed. Yalow's method is used for a great variety of purposes and earned her the 1977 Nobel Prize in physiology or medicine.

Yang, Chen Ning (b. 1922) Chinese-born American physicist who, in collaboration with Tsung Dao Lee, demonstrated that parity is not conserved by the weak nuclear force. Parity is the property of a wave function that determines its behavior when all its spatial coordinates are reversed in direction. It is conserved in electromagnetic and strong interactions. This was a critical advance in the development of theoretical physics and it earned Yang and Lee the Nobel Prize in physics in 1957.

Young, Thomas (1773–1829) English polymath many of whose achievements in physics were so far ahead of his time that they were ignored in his lifetime. He studied medicine; explained eye focusing, astigmatism, and color vision; explained the physics of elasticity; established the wave theory of light; and translated the Rosetta stone and elucidated Egyptian hieroglyphics. His notes remained unknown and neglected in the archives of the Royal Institution in London until after many of his discoveries had been made again by others. He had been forced to resign his professorship at the Royal Institution because his lectures were much too difficult for a popular audience to understand.

Yukawa, Hideki (1907–81) Japanese physicist who first described the strong nuclear force. This is the force that holds the nuclear protons and neutrons together in spite of the fact that like charges repel each other. It is about 100 times the strength of electromagnetic forces. Yukawa suggested that the strong force involved the exchange of a particle and described its characteristics. Such a particle, the pi-meson or pion, was discovered in 1947. Yukawa was awarded the Nobel Prize in physics in 1949.

Pieter Zeeman

Zeeman, Pieter (1865–1943) Dutch physicist who showed that the spectral lines of light from sodium and lithium are split into two or three by a magnetic field. This is known as the Zeeman effect and is the result of the fact that atoms emit several closely spaced lines, each originating in a particular quantum energy level. For this work, Zeeman shared the 1902 Nobel Prize in physics with Lorentz.

SECTION THREE CHRONOLOGY

c. 6000 BCE • Clay tokens in use to record numbers and measures for trade transactions in Anatolia

c. 2500 BCE • Standardized weights used in Sumer

c. 1750 BCE • Mathematical and geometric knowledge recorded in Egyptian papyrus known as Moscow Papyrus

c. 600 BCE • Earliest known proof of what later becomes known as Pythagorean theorem contained in Chinese mathematics text

c. 580 BCE • Early theory of matter proposed by Greek philosopher Thales, suggesting that all things are made from forms of water

c. 570 BCE • Earliest proofs of geometric propositions given by Greek philosopher Thales

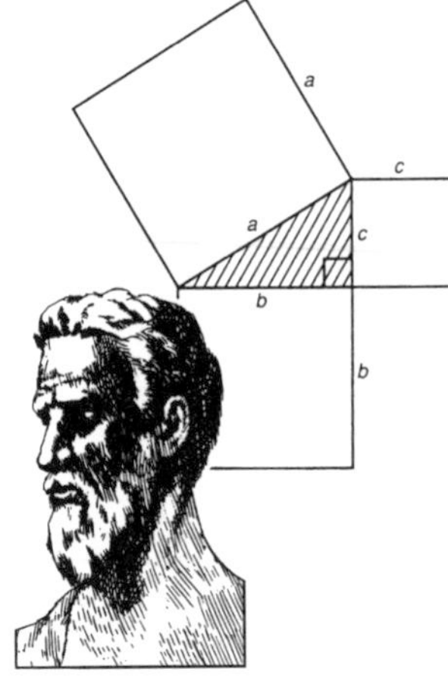

c. 550 BCE
Pythagorean Brotherhood makes discoveries in mathematics.

c. 550 BCE • Pythagorean Brotherhood in southern Italy makes fundamental discoveries in mathematics

c. 540 BCE • Matter explained in terms of cold, heat, dryness, and wetness by Greek philosopher Anaximander

c. 530 BCE • Theory that primary matter is made of air proposed by Greek philosopher Anaximines

480–471 BCE • Greek philosopher Anaxagoras suggests materials are made up of large numbers of "seeds," which combine in different proportions to make different materials

c. 445 BCE • Atomic theory of matter introduced by Greek philosopher Leucippus

440 BCE • Four-element theory of matter (fire, air, water, earth) introduced by Greek philosopher Empedocles

430–421 BCE • Concept of atoms is expanded by Greek philosopher Democritus

c. 350 BCE • Free fall is found to be an accelerated motion by Greek philosopher Aristotle

350–341 BCE • Greek philosopher Aristotle defines chemical elements as constituents of bodies that cannot be decomposed into other constituents

340–331 BCE • Greek physicist Strato believes bodies accelerate when they fall

310–301 BCE • Greek philosopher Epicurus founds philosophical school based on a theory of atoms

c. 300 BCE • Greek mathematical knowledge summarized by Greek mathematician Euclid

c. 260 BCE • Lever power demonstrated by Greek mathematician Archimedes. Archimedes' screw for raising water from one level to another invented

c. 250 BCE • Accurate estimate of Earth's circumference made by Greek mathematician Eratosthenes

c. 225 BCE • Multiple pulley system to allow large loads to be lifted invented by Greek mathematician Archimedes

c. 95 BCE • Differential gearing invented in Greece

60 BCE • Roman philosopher Lucretius speculates that matter is made up of atoms

97 CE • Existing engineering knowledge compiled in two-volume work on aqueducts by Roman Frontinus

c. 185 • Earliest known work on alchemy, forerunner of chemistry, compiled in Egypt

c. 250 • Earliest known work on algebra compiled by Roman mathematician Diophantus

400–409 • Term *chemistry* used for first time by Alexandrian scholars to describe the activity of changing matter

c. 500 • Abacus in use in Europe as an arithmetic aid

c. 600 • Decimal notation devised by Indian mathematicians Aryabhata and Varamihara

789 • Standard units of weight and measure introduced by Charlemagne, Frankish king (later emperor of the West)

876 • First known use of the symbol for zero featured in Hindu inscriptions

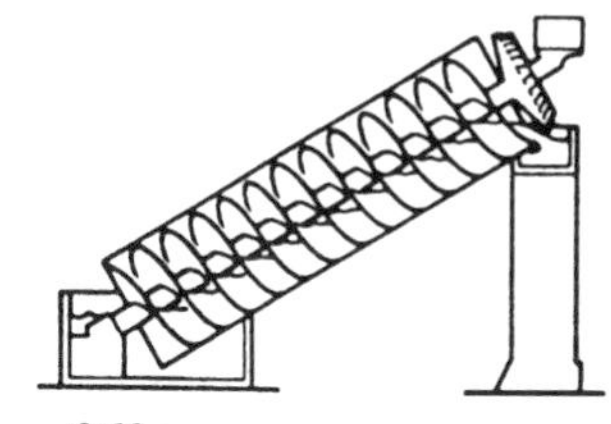

c. 260 BCE
Archimedes' screw invented.

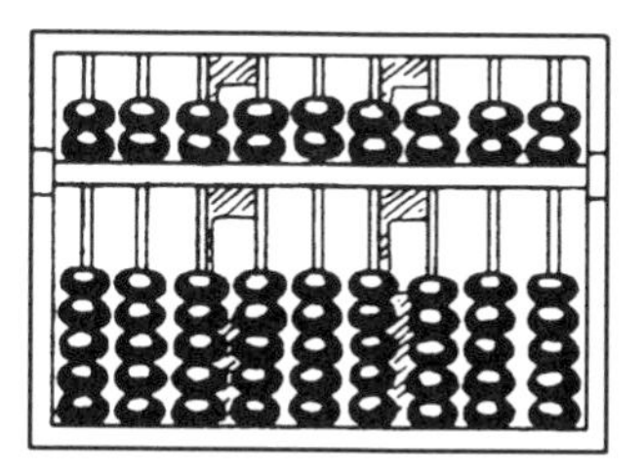

c. 500
Abacus in use in Europe.

c. 880 • Chemical substances classified by Persian alchemist Rhazes into animal, vegetable, mineral, and derivative; minerals classified as salts, spirits, and stones

c. 1010 • Correct account of how lenses work written by Arab physician Alhazen; he also develops parabolic mirrors

c. 1190 • Magnetic compasses used in Europe

1250 • Magnifying glass invented by English philosopher Roger Bacon. Element arsenic discovered by German scientist Albertus Magnus

1269 • First scientific account of magnetic poles and the compass

1270 • Polish natural scientist Witelo produces treatise on optics dealing with refraction, reflection, and geometrical optics

c. 1300 • Sulfuric acid described. Eyeglasses in widespread use

c. 1350 • Concept of impetus developed by French philosopher Jean Buridan to explain motion of heavenly bodies

1490 • Capillary action of liquids in narrow tubes noted by Italian artist Leonardo da Vinci

1492 • Italian artist Leonardo da Vinci studies concept of friction. Italian explorer Christopher Columbus discovers that a magnetic compass changes the direction in which it points as longitude changes

1514 • Earliest use of "+" and "–" signs in mathematics by Dutch mathematician Vander Hoecke

1543 • Aristotle's theories of motion and space attacked by French scientist Petrus Ramus

1583 • Principle of pendulum discovered by Italian scientist Galileo Galilei. Science of hydrostatics founded by Dutch mathematician Simon Stevinus

1590 • Aristotle's theories of motion refuted by Italian scientist Galileo Galilei

1592 • Early thermometer invented by Italian scientist Galileo Galilei

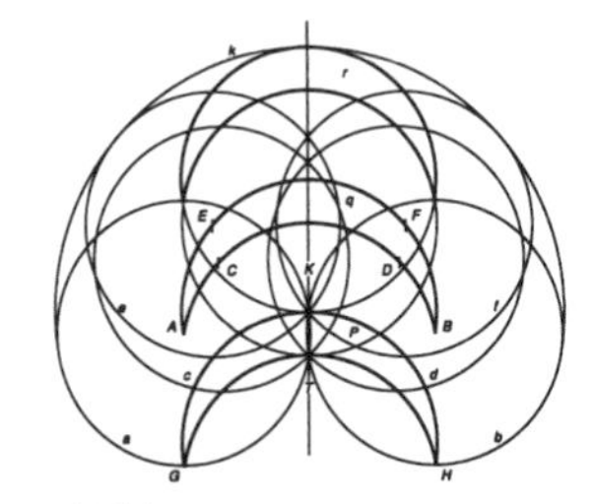

c. 1010
Alhazen develops parabolic mirrors.

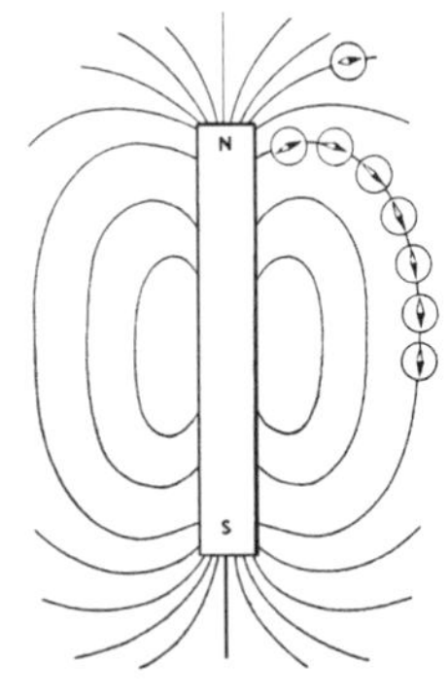

1492
Pattern of lines of force about a magnet, the basis for a magnetic compass.

1600 • Discovery of elements antimony and bismuth. Theory that Earth is a huge magnet put forward by English physician William Gilbert, who also studies static electricity

1604 • Law of free fall (that it is a constant for all bodies) announced by Italian scientist Galileo Galilei following experiments carried out from the Leaning Tower of Pisa

1608 • Refracting telescope patented by Dutch instrument maker Hans Lippershey

1614 • Logarithms described by Scottish mathematician John Napier

1620 • Term *gas* coined by Flemish chemist Jean van Helmont to describe airlike substances

1621 • Snell's law concerning refraction formulated by Dutch mathematician Willebrod Snell

1622 • Based on his study of logarithms, English mathematician William Oughtred invents one of the first slide rules

1630 • Term *lines of force* first used by Italian natural philosopher Niccolò Cabeo

1631 • Sign "x" for multiplication introduced by English mathematician William Oughtred

1637 • Philosophical study of scientific method (*Discours de la méthode*) published by French philosopher René Descartes, who also publishes theory of refraction

1640 • Science of hydrodynamics founded by Italian physicist Evangelista Torricelli. Micrometer invented by English scientist William Gascoigne

1647 • Calculating machine invented by French mathematician Blaise Pascal

1648 • Pascal's law, concerning the effect of applying pressure to a fluid in a closed container, formulated by French mathematician Blaise Pascal

1649 • Atomic theory of matter revived by French philosopher Pierre Gassendi

1614
John Napier describes logarithms.

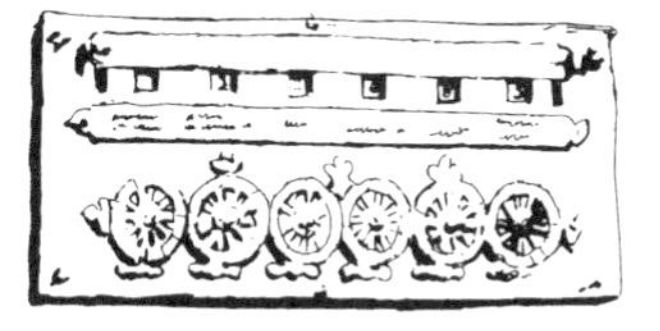

1647
Blaise Pascal invents calculating machine.

1654 • Calculus pioneered by Swiss mathematician Jakob Bernouilli

1659 • German mathematician Johann Heinrich Rahn introduces "÷" sign for division

1660 • Concepts of elements, acids, and alkalis formulated by Irish chemist Robert Boyle. Existence of static electricity demonstrated by German physicist Otto von Guericke

1661 • Dutch physicist Christiaan Huygens invents manometer

1662 • Boyle's law, on the relationship between pressure and the volume of a gas, stated by Irish chemist Robert Boyle

1665 • English physicist Isaac Newton describes composition of white light and invents a form of calculus

1668 • Reflecting telescope invented by English physicist Isaac Newton and used by him to study a wide range of celestial bodies and phenomena

1668
Isaac Newton invents reflecting telescope.

1669 • Danish physician Erasmus Bartholin observes phenomenon of double refraction. Element phosphorus discovered by German chemist Hennig Brand

1675 • Danish astronomer Olaus Roemer estimates speed of light as 141,000 miles (227,000 km) per second; about 75% of its true speed

1676 • English chemist Robert Hooke formulates Hooke's law on elastic bodies

1678 • Dutch physicist Christiaan Huygens maintains that light consists of waves

1679 • Idea that movement of planets is a purely mechanical problem solvable by mathematics first put forward by English chemist and astronomer Robert Hooke

1687 • In his *Philosophiae naturalis principia mathematica*, English physicist Isaac Newton publishes his three laws of motion and law of universal gravitation

1700 • Term *acoustics*, with reference to the behavior of sound, coined by French physicist Joseph Sauveur

1704 • Scientific and mathematical study of optics, suggesting that light is made of particles, published by English physicist Isaac Newton

1705 • Demonstration that sound cannot travel in a vacuum carried out by English scientist Francis Hauksbee

1706 • Greek letter "π" (pi) used as symbol for ratio of a circle's circumference to its diameter for the first time

1714 • Mercury thermometer invented by German physicist Gabriel Fahrenheit

1723 • First known treatise on the study of crystallography published by M. A. Capeller

1729 • Dutch physicist Pieter van Musschenbroek becomes one of the first to use the term *physics*. Electricity found to travel through conductors by English physicist Stephen Gray

1730 • Alcohol thermometer developed by French physicist René de Réaumur

1733 • Two types of static electricity distinguished by French scientist Charles Du Fay

1735 • Discovery of element platinum by Spanish scientist Antonio de Ulloa. Discovery of element cobalt by Swedish chemist Georg Brandt

1738 • Lift (Bernoulli effect – later used in airfoil) discovered by Swiss physicist Daniel Bernoulli

1742 • Celsius temperature scale devised by Swedish astronomer Anders Celsius, although with boiling point at zero and freezing point at 100 degrees

1743 • Formulation of d'Alembert principle by French physicist Jean le Rond d'Alembert

1744 • Nature of heat as a form of motion described by Russian scientist Mikhail Lomonosov

1745 • Independent invention of Leyden jar by Dutch physicist Pieter van Musschenbroek and German physicist Ewald Georg von Kleist

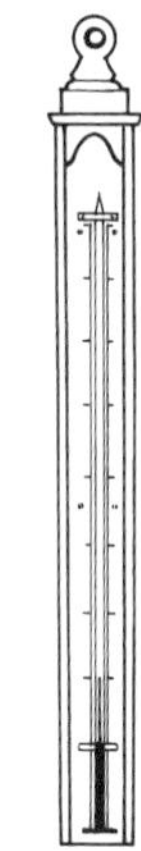

1714
Mercury thermometer.

1742
Stamp with Celsius portrait.

1748 • Osmotic pressure (concerning solutions) discovered by French physicist Jean Nollet

1751 • Element nickel isolated by Swedish mineralogist Axel F. Cronstedt. Positive and negative electricity distinguished by American scientist Benjamin Franklin

1752
Benjamin Franklin proves lightning is a form of electricity.

1752 • Lightning shown to be electrical by American scientist Benjamin Franklin. Some principles of hydrodynamics formulated by French physicist Jean le Rond d'Alembert

1754–56 • Scottish chemist Joseph Black shows that carbon dioxide is different from ordinary air

1761 • Latent heat discovered by Scottish chemist Joseph Black

1766 • Element hydrogen discovered by English chemist Henry Cavendish

1771 • Element fluorine discovered by Swedish chemist Carl Wilhelm Scheele

1772 • Element nitrogen discovered independently by Scottish chemist Daniel Rutherford, English chemist Joseph Priestley, and two other chemists. Element oxygen discovered by German chemist Carl Wilhelm Scheele. French scientist Jean Romé de Lisle describes process of crystallization

1774 • Discovery of element manganese credited to Swedish mineralogist Johan Gottlieb Gahn

1775 • Improved anemometer (to measure wind speed) invented by Irish physicist James Lind

1777 • Invention of torsion balance by French physicist Charles Augustin de Coulomb

1778 • Discovery of element molybdenum by Swedish chemist Carl Wilhelm Scheele

1779 • Discovery of laws of friction by French physicist Charles Augustin de Coulomb

1780 • Calorimeter developed by French chemists Pierre Simon Laplace and Antoine Lavoisier

1784 • English physicist George Atwood determines the acceleration of a free-falling body. Discovery of element tellurium by Austrian mineralogist Franz Joseph Müller

1785 • Coulomb's law formulated by French physicist Charles Augustin de Coulomb

1788 • Formulation of Lagrangian function by French mathematician Joseph Louis Lagrange

1789 • A compound of uranium discovered by German chemist Martin Heinrich Klaproth

1790 • Irish physician Adair Crawford discovers element strontium

1791 • Discovery of element titanium by English chemist William Gregor

1794 • Finnish chemist Johan Gadolin discovers element yttrium

1797 • Discovery of element chromium by French chemist Louis-Nicolas Vauquelin

1798 • Discovery of friction-generated heat by Anglo-American physicist Benjamin Thompson, Count Rumford. Gravitational constant determined by English chemist Henry Cavendish. Discovery of element beryllium by French chemist Louis-Nicolas Vauquelin

1800 • English chemists William Nicholson and Anthony Carlisle use electricity to produce chemical change. Infrared radiation discovered by German-English astronomer William Herschel. Voltaic pile for storing electricity (the first battery) devised by Italian physicist Alessandro Volta

1801 • Element niobium discovered by English chemist Charles Hatchett. English chemist John Dalton formulates law of partial pressures of gases (Dalton's law). Henry's law, that the amount of gas absorbed by a liquid varies directly with the pressure, formulated by English chemist and physicist William Henry. Ultraviolet radiation discovered by German physicist Johann Ritter

1802 • Discovery of element tantalum by Swedish chemist Anders G. Ekeberg. English chemist John Dalton makes atomic weight tables

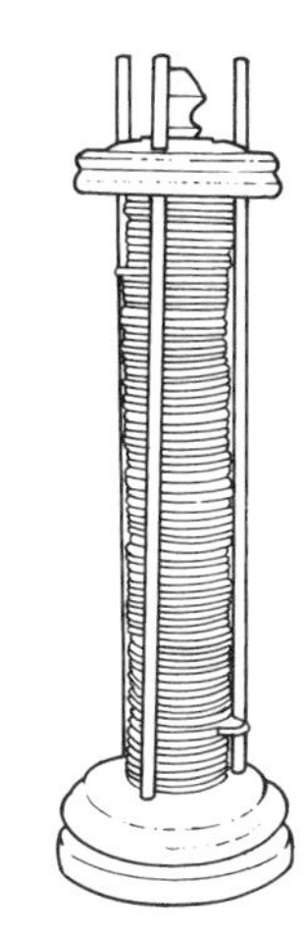

1800
Alessandro Volta uses "piles" of alternating zinc and silver disks in the first battery.

1803
John Dalton proposes that matter comprises atoms.

1803 • Discovery of elements palladium and rhodium by English chemist William H. Wollaston. Discovery of elements iridium and osmium by English chemist Smithson Tennant. Discovery of element cerium by Swedish chemist Jons Jakob Berzelius, Swedish mineralogist Wilhelm Hisinger, and German chemist Martin H. Klaproth. Theory that matter is made up of atoms proposed by English chemist John Dalton

1804 • Principles of aerodynamics described by English engineer George Cayley

1805 • Composition of water (hydrogen and oxygen) established by French chemist Joseph Gay-Lussac

1806 • Wind scale devised by British naval officer Francis Beaufort

1807 • Discovery of elements sodium and potassium by English chemist Humphry Davy. Element boron isolated by Davy. English physicist Thomas Young is first to use the word *energy* in a sense that is similar to modern usage

1808 • Elements barium, magnesium, and calcium discovered by English chemist Humphry Davy

1810 • English chemist Humphry Davy credited with identification of chlorine as an element. English chemist John Dalton formulates atomic weight theory

1811 • Brewster's law formulated by Scottish physicist David Brewster. Term *molecule* first used by Italian physicist Amedeo Avogadro. Discovery of element iodine by French chemist Bernard Courtois

1817 • Discovery of element lithium by Swedish chemist Johan August Arfwedson. Discovery of element cadmium by German chemist Friedrich Strohmeyer

1818 • Discovery of element selenium by Swedish chemist Jons Jakob Berzelius. Process of light diffraction discovered by French physicist Augustin Fresnel

1820 • Diffraction grating invented by German physicist Joseph von Fraunhofer. Electromagnetism discovered by Danish physicist Hans Christian Oersted

1822 • First known observation of thermoelectricity observed by Estonian-German physicist Thomas Seebeck. Principle and first model of electric motor devised by English physicist Michael Faraday

c. 1824 • Discovery of element silicon by Swedish chemist Jons Jakob Berzelius

1825 • Discovery of element aluminum by Danish physicist Hans Christian Oersted. First practical electromagnet built by English scientist William Sturgeon. First moving-coil galvanometer developed by German physicist Johann Schweigger

1826 • Element bromine discovered by French chemist Antoine-Jérôme Balard

1827 • Brownian motion discovered by Scottish botanist Robert Brown. Ohm's law, relating to electrical resistance, formulated by German physicist Georg Simon Ohm

1828 • Discovery of element thorium by Swedish chemist Jons Jakob Berzelius

1829 • Term *kinetic energy* coined by French physicist Gaspard Gustave Coriolis. Graham's law on the diffusion rate of gases formulated by Scottish scientist Thomas Graham. Nicol prism invented by Scottish physicist William Nicol

1830 • Discovery of element vanadium by Swedish chemist Nils G. Sefström. Principle of electromagnetic induction described by American physicist Joseph Henry

1831 • English physicist Michael Faraday invents induction coil

1833 • Stroboscope invented by Austrian scientist Simon von Stampfer. Differential calculating machine invented by English mathematician Charles Babbage. Terms *electrode*, *cathode*, *anode*, *ion*, *cation*, *anion*, *electrolyte*, and *electrolysis* coined by English physicists Michael Faraday and William Whewell

1834 • Peltier effect (on the thermoelectric reduction of temperature) discovered by French physicist Jean Charles Athanase Peltier

1831
Michael Faraday discovers electromagnetic induction.

1839 • Element lanthanum discovered by Swedish chemist Carl Gustaf Mosander

1840 • First of Joule's laws formulated by English physicist James P. Joule

1842 • Austrian physicist Christian Doppler discovers Doppler effect

1843 • Mechanical equivalent of heat formulated by English physicist James P. Joule. Discovery of elements erbium and terbium by Swedish chemist Carl Gustaf Mosander

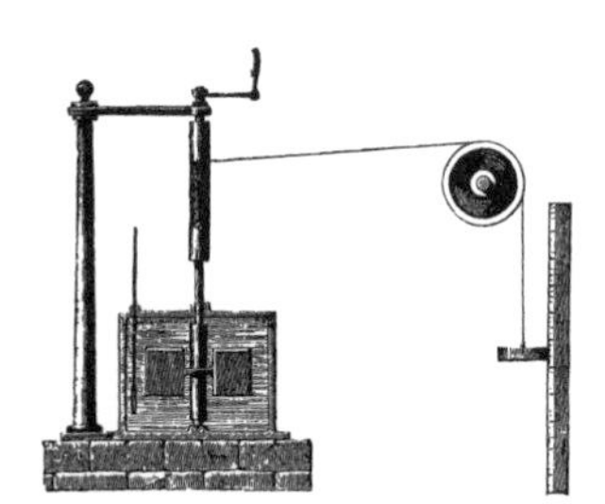

1843
James Joule devises a method of measuring mechanical equivalent of heat.

1844 • Discovery of the pure element ruthenium by Russian chemist Karl K. Klaus

1845 • Paramagnetism and diamagnetism discovered by English physicist Michael Faraday

1847 • First law of thermodynamics stated by German physicist Hermann Ludwig Ferdinand von Helmholtz

1848 • Concept of absolute zero formulated by Scottish physicist William Thomson (later Lord Kelvin)

1849 • Term *thermodynamics* introduced by Scottish physicist William Thomson (later Lord Kelvin)

1850 • First true transformer developed by German instrument maker Heinrich Ruhmkorff. Second law of thermodynamics formulated by German physicist Rudolf Julius Emanuel Clausius

1852 • Joule-Thomson effect demonstrated by English physicist James P. Joule and Scottish physicist William Thomson (later Lord Kelvin). Term *fluorescence* coined by British physicist George Stokes

1853 • Concept of potential energy developed by Scottish physicist William J. M. Rankine. Light shown to travel faster in air than in water by French physicist Léon Foucault

1855 • Ruhmkorff induction coil invented by German physicist Heinrich Ruhmkorff

1859 • Kinetic theory of gases developed by Scottish physicist James Clerk Maxwell

1860 • Element cesium discovered by German chemists Robert Bunsen and Gustav Kirchhoff. Statistics for analyzing the behavior of molecules in a gas independently developed by Scottish physicist James Clerk Maxwell and Austrian physicist Ludwig Boltzmann

1861 • Element rubidium discovered by German chemists Robert Bunsen and Gustav Kirchhoff. Element thallium discovered by English physicist William Crookes

1863 • Discovery of element indium by German mineralogists Ferdinand Reich and Theodor Richter

1865 • Maxwell's equations formulated by Scottish physicist James Clerk Maxwell

1868 • Discovery of element helium by French astronomer Pierre-Jules-César Janssen and English astronomer Joseph Norman Lockyer

1869 • First periodic table for chemical elements published by Russian chemist Dmitri Mendeleyev. Tyndall effect discovered by Irish physicist John Tyndall

1873 • Basic laws of electromagnetism formulated by Scottish physicist James Clerk Maxwell

1875 • Element gallium discovered by French chemist Paul-Émile Lecoq de Boisbaudran. Invention of radiometer by English physicist William Crookes. Discovery of Kerr effect (magneto-optic) by Scottish physicist John Kerr

1876 • Cathode rays discovered by German physicist Eugen Goldstein

1878 • Element holmium discovered by Swiss scientists J. L. Soret and Marc Delafontaine

1879 • Element scandium discovered by Swedish chemist Lars Nilson. Element samarium discovered by French chemist Paul-Émile Lecoq de Boisbaudran. Element thulium discovered by Swedish chemist Per Cleve. Hall effect discovered by Edwin H. Hall. Austrian physicist Josef Stefan formulates Stefan's law. Practical electric light with long-lasting filament and gas-filled bulb invented

1873
James Clerk Maxwell formulates laws of electromagnetism.

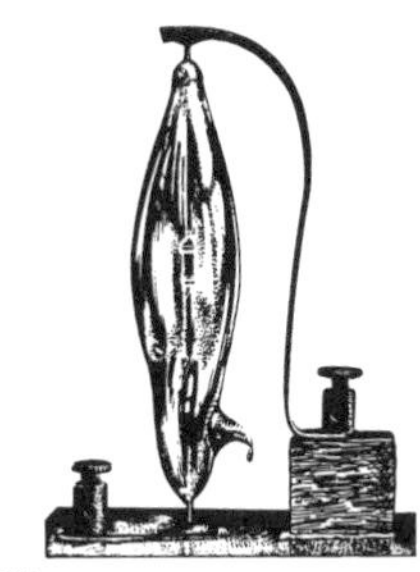

1879
Joseph Swan and Thomas Edison invent the electric light.

independently by American inventor Thomas Edison and English physicist Joseph Swan

1880 • Element gadolinium discovered by Swiss chemist Jean-Charles de Marignac. Piezoelectricity discovered by French chemist Pierre Curie

1881 • Concept of electromagnetic mass introduced by English physicist Joseph John Thomson

1883 • Edison effect discovered by American inventor Thomas Edison

1883
Thomas Edison discovers Edison effect.

1885 • Discovery of elements neodymium and praseodymium by Austrian chemist Carl Auer

1886 • Channel (or canal) rays discovered by German physicist Eugen Goldstein. Element germanium discovered by German chemist Clemens Winkler. Element dysprosium discovered by French chemist Paul-Émile Lecoq de Boisbaudran

1887 • Photoelectric effect discovered by German physicist Heinrich Rudolf Hertz

1892 • FitzGerald contraction of bodies in motion suggested by Irish physicist George Francis FitzGerald

1893 • American physicist Albert A. Michelson invents interferometer for measuring wavelength of light

1894 • Element argon discovered by Scottish chemist William Ramsay and English scientist Lord Rayleigh

1895 • Cloud chamber developed by Scottish physicist Charles Thomson Rees Wilson. Lorentz force discovered by Dutch physicist Hendrik Lorentz. Scottish chemist William Ramsay first to discover element helium on Earth

1896 • Zeeman effect discovered by Dutch physicist Pieter Zeeman

1897 • Electron discovered and its mass calculated by English physicist Joseph John Thomson

1897
Joseph John Thomson discovers electron.

1898 • Elements radium and polonium discovered by French chemist Pierre Curie and French physicist Marie Curie. Term *radioactivity* coined by Marie Curie. Discovery of elements krypton, neon, and xenon by Scottish chemist William Ramsay and English chemist Morris William Travers

1899 • Alpha and beta rays, two types of radioactivity, distinguished by British physicist Ernest Rutherford. Discovery of element actinium by French chemist André-Louis Debierne

1900 • British physicist Ernest Rutherford names beta rays (later known as beta particles). English physicist Owen Willans Richardson discovers that heated metals tend to emit electrons. Planck's radiation law stated by German physicist Max Ernst Ludwig Planck. Element radon discovered by German physicist Friedrich Ernst Dorn

1900
Max Ernst Ludwig Planck develops law of radiation.

1901 • Discovery of element europium by French chemist Eugène-Anatole Demarçay

1903 • Chaos theory formulated by French physicist Jules Henri Poincaré

1905 • Special theory of relativity devised by German-born physicist Albert Einstein, including formula $E = mc^2$

1907 • Element lutetium discovered by French chemist Georges Urbain. Element ytterbium discovered by Swiss chemist Jean-Charles de Marignac

1908 • Paschen series of lines discovered by physicist Louis Paschen. International ampere adopted as basic unit of electric current

1910 • Existence of isotopes confirmed by English physicist Joseph John Thomson

1913 • Discovery of Stark effect by German physicist Johannes Stark. English chemist Frederick Soddy coins the term *isotope*. Quantum theory proposed by Danish physicist Niels Bohr

1914 • Name *proton* given to the positively charged nucleus of the hydrogen atom by British physicist Ernest Rutherford

1915 • General theory of relativity conceived by German-born physicist Albert Einstein. Sonar (*so*und *na*vigation and *r*anging) invented by French physicist Paul Langevin

1919
Ernest Rutherford first to produce atomic fission artificially.

1917 • Discovery of element protactinium by German chemist Otto Hahn and Austrian physicist Lise Meitner

1919 • First artificial atomic fission achieved by British physicist Ernest Rutherford

1921 • Half-integer quantum numbers are discovered by German physicist Alfred Landé

1923 • American physicist Arthur Holly Compton coins word *photon*. Element hafnium discovered by Dutch physicist Dirk Coster and Hungarian chemist George de Hevesy

1924 • Discovery of wave nature of electrons by French physicist Louis Victor de Broglie. Frequency modulation (FM) devised by American electrical engineer Edwin Armstrong. Indian physicist Satyendra Nath Bose and German-born physicist Albert Einstein colaborate to develop Bose-Einstein statistics

1925 • Element rhenium discovered by German chemists Walter Noddack, Ida Tacke, and Otto C. Berg. First analog computer built by American electrical engineer Vannevar Bush. Matrix mathematics developed by German physicist Werner Karl Heisenberg for studying energy levels of electrons. Packing fraction discovered by English physicist Francis Aston. Auger effect discovered by Pierre Auger

1926 • Invention of Bernal chart by Irish physicist John Desmond Bernal to assist with the analysis of crystal structures

1927 • Electron diffraction of crystals independently discovered by American physicist Clinton Joseph Davisson and English physicist George Paget Thomson

1928 • Concept of parity of atomic states developed by Hungarian-American physicist Eugene Paul Wigner

1929 • German-born physicist Albert Einstein develops a unified field theory

1930 • Discovery of superfluidity

1931 • Concept of resonance developed by American chemist Linus Carl Pauling. Electron microscope invented by German scientists Ernst Ruska and Max Knoll

1932 • First nuclear reaction to result from the bombardment of an element by artificially accelerated particles achieved by English physicist John Cockcroft and Irish physicist Ernest Walton. Neutron discovered by English physicist James Chadwick

1933 • Magnetic characteristics and wave aspects of molecular beams demonstrated by German-American physicist Otto Stern

1934 • Mass of a neutron determined by English physicist James Chadwick and Austrian-American physicist Maurice Goldhaber. Cherenkov radiation discovered by Russian physicist Pavel Alekseyevich Cherenkov. Artificial radioactivity achieved for first time by French physicists Irène and Frédéric Joliot-Curie. Nucleus of atom split by American physicist Enrico Fermi

1935 • Radar (radio detection and ranging) developed by Scottish physicist Robert Watson-Watt. American seismologist Charles Richter develops Richter scale for measuring the intensity of earthquakes

1937 • Concept of charge conjugation developed by physicist H. A. Kramers. "Muon" particle discovered by American physicist Carl David Anderson. Element technetium discovered by French physicist Carlo Perrier and Italian-born American Emilio Segrè

1938 • Discovery of chain reaction nuclear fission by German chemist Otto Hahn and Austrian physicist Lise Meitner. Technique of magnetic resonance developed by American physicist Isidor Isaac Rabi

1939 • Method for calculating properties of substances from quantum principles discovered by W. C. Herring. Magnetic movement of a neutron calculated by American physicist Felix Bloch. Discovery of element francium by French physicist Marguerite Perey

1940 • Discovery of element astatine by American physicists Emilio Segrè, Dale Corson, and K. R. Mackenzie. Discovery of element plutonium by American physicist Glenn T. Seaborg and colleagues. Discovery of element neptunium

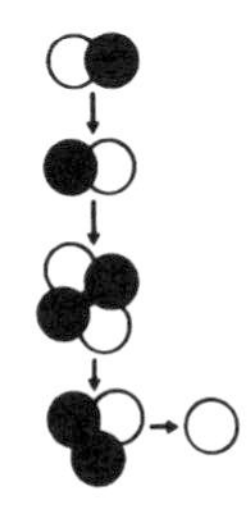

1932
James Chadwick discovers neutron.

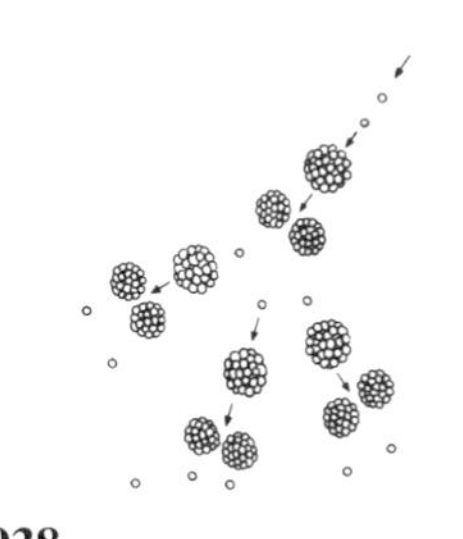

1938
Bombarding uranium with neutrons creates a chain reaction.

by American physicists Edwin M. McMillan and Phillip H. Abelson

1942 • First controlled chain reaction in a uranium and graphite pile created by American physicist Enrico Fermi, pointing the way to industrial production of nuclear power

1944 • Discovery of elements americium and curium by American physicist Glenn T. Seaborg and colleagues

1946 • Development of nuclear magnetic resonance (NMR) technique by American physicists Felix Bloch and Edward Mills Purcell

1947 • Theory of holography developed by Hungarian-British physicist Dennis Gabor. First true meson (pi-meson or pion) discovered by English physicist Cecil Frank Powell. Element promethium discovered by American chemists J. A. Marinsky, L. E. Glendenin, and C. D. Coryell

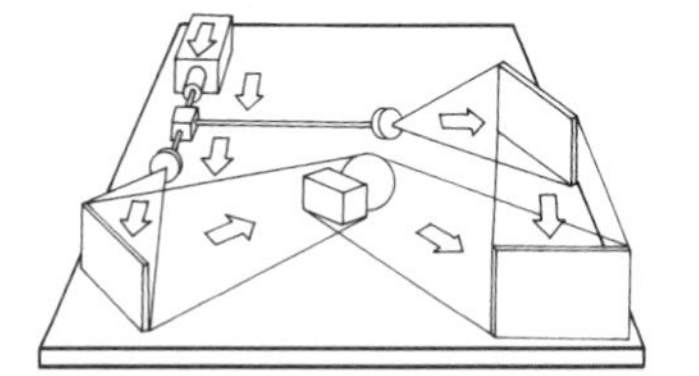

1947
Recording a transmission hologram.

1948 • Theory of quantum electrodynamics (QED) developed by American physicist Richard Phillips Feynman. Shell model of atomic nucleus advanced by American physicist Maria Goeppert-Mayer and German physicist Johannes Hans Daniel, who independently introduce concept of magic numbers. Transistor invented by American physicists William Shockley, Walter Brattain, and John Bardeen

1949 • Discovery of element berkelium by American physicist Glenn T. Seaborg and colleagues

1950 • Discovery of element californium by American physicist Glenn T. Seaborg and colleagues

1950s • Two main classes of elementary particles are discovered: hadrons (including nucleons, mesons, and hyperons) and leptons (including electrons, neutrinos, and muons)

1952 • Discovery of K meson (or kaon) and lambda particle by Polish physicists Marian Danysz and Jerzy Pniewski. Discovery of element einsteinium by American physicist Albert Ghiorso and colleagues. Discovery of element fermium by American physicist Albert Ghiorso and colleagues

1953 • Bubble chamber, for detecting ionizing radiation, invented by American physicist Donald Arthur Glaser. Maser (microwave amplification by stimulated emission of radiation), forerunner of laser, produced by American physicist Charles Townes

1954 • First particle accelerator built

1955 • Invention of field ion microscope, the first device to yield images of individual atoms, by American physicist Erwin Wilhelm Mueller. Two types of K mesons detected with differing modes of decay. Discovery of element mendelevium by American physicist Albert Ghiorso and colleagues. Antiprotons discovered; they are negatively charged particles that have the mass of protons

1956 • American physicists Frederick Reines and Clyde Larrain Cowan discover antineutrinos

1958 • Discovery of element nobelium by American physicist Albert Ghiorso and colleagues

1959 • Japanese physicists Saburo Fukui and Shotaro Miyamoto invent spark chamber to detect ionizing particles selectively

1960 • Laser (light amplification by stimulated emission radiation) constructed by American physicist Theodore Maiman

1961 • Record set by Russian military scientists for largest nuclear explosion when they test a 58-megaton weapon. Elementary particles called hadrons classified by American physicist Murray Gell-Mann in a system he calls Eightfold Way. American physicist Albert Ghiorso and colleagues discover element lawrencium

1962 • British physicist Heinz London develops technique for inducing very low temperatures with mixture of helium-3 and helium-4. Canadian chemist Neil Bartlett combines noble gas xenon with platinum fluoride to produce xenon fluoroplatinate; first known case of a noble gas bonding with another element to form a compound

1967 • Discovery of element dubnium by Russian scientists

1969 • Discovery of element rutherfordium by American physicist Albert Ghiorso and colleagues

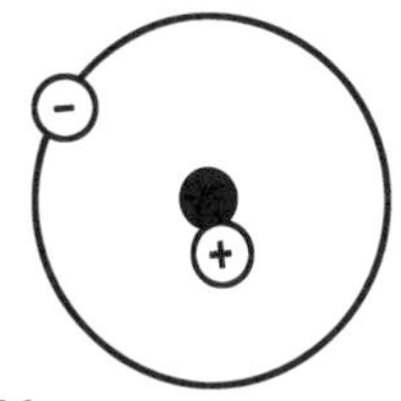

1956
Discovery of antineutrinos.

1961
Russians explode largest nuclear bomb.

1970 • Invention of scanning electron microscope by American physicist Albert Victor Crewe. Element hahnium discovered by American physicists Albert Ghiorso and colleagues

1972 • American physicist Kenneth M. Evenson and colleagues obtain new level of precision in measuring speed of light: 186,282.3959 miles (299,172.77 km) per second

1974 • Element seaborgium discovered by American physicists Albert Ghiorso and colleagues

1980 • Development of scanning tunneling microscope, which can produce images of individual atoms on the surface of material

1981 • Physicists in Darmstadt, Germany, confirm the existence of element bohrium, after Russian scientists originally report its discovery in 1976

1982 • Discovery of element meitnerium by German scientists

1984 • Discovery of first quasi-crystal by American scientist D. Schechtman. German scientists discover element hassium

1985 • Lanxides, substances with the properties of both metal and ceramics, are discovered by American researchers

1989 • Mathematicians extend calculation of pi to 1 billion digits

1990 • Existing records broken when two computer scientists factor a 155-digit Fermat number

1991 • American chemist Joel Hawkins corroborates existence of buckyball molecule (or buckminsterfullerene), a form of pure carbon

1992 • British scientist David Slowinski discovers largest prime number to date; it is 227,832 digits long

1993 • Fermat's last theorem reportedly proved by British mathematician Andrew Wiles

1993
Andrew Wiles proves Fermat's last theorem.

SECTION FOUR
CHARTS & TABLES

Table of physical quantities

Quantity	Symbol	SI unit	Unit symbol
length	l, L, x	meter	m
mass	M, m	kilogram	kg
time	t, T	second	s
temperature	t, T, θ	kelvin	K
energy	Q, u	joule	J
pressure	P, p	pascal	Pa
electric charge	Q, q	coulomb	C
electric current	I, i	ampere	A
potential difference	V	volt	V
electrical resistance	R	ohm	n
magnetic field intensity	B	tesla	T

Constant	Symbol	Si value
Avogadro's number	N_A	6.025×10^{23} gmole^{-1}
Planck's constant	h	6.625×10^{-34} Js
free space speed of light	c	3.00×10^{8}ms^{-1}
electron charge	e	1.602×10^{-19} C
electron rest mass	m_e	9.11×10^{-31} Ckg
specific electron charge	e/m	1.760×10^{11} Ckg^{-1}
atomic mass unit	amu	1.660×10^{-27} kg
proton rest mass	m_p	1.6724×10^{-27} kg
neutron rest mass	m_n	1.6733×10^{-27} kg
Stefan-Boltzmann constant	a	5.67×10^{-8} Jm^{-2}K^{-4}s^{-1}
universal gas constant	R	8.31JK^{-1} g mole^{-1}
universal gravitation constant	G	6.673×10^{-11} Nm^{2}kg^{-2}
Boltzmann constant	k	1.381×10^{-23} JK^{-1}

Intensity and loudness

Sound intensity in decibels (dB) increases with the energy of sound-producing vibrations as these increase sound waves' amplitude. Loudness is perceived intensity: mid-frequency sounds seem louder than high or low sounds of the same intensity. The chart and scale illustrate common noises and decibel ratings at certain distances. Noises at 120–130 dB cause pain: above 140 dB they can harm ears permanently.

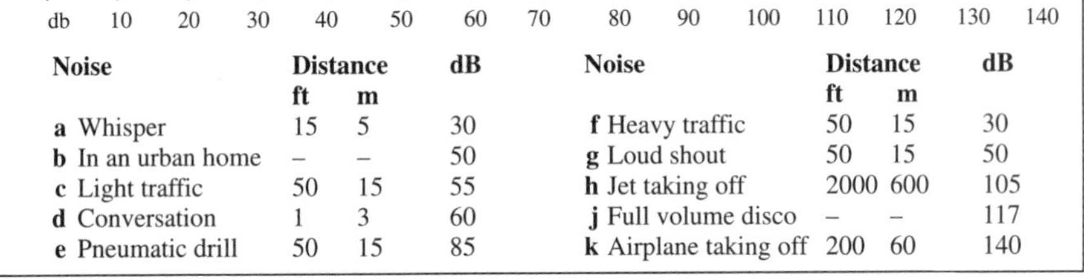

Noise	Distance ft	Distance m	dB
a Whisper	15	5	30
b In an urban home	–	–	50
c Light traffic	50	15	55
d Conversation	1	3	60
e Pneumatic drill	50	15	85
f Heavy traffic	50	15	30
g Loud shout	50	15	50
h Jet taking off	2000	600	105
j Full volume disco	–	–	117
k Airplane taking off	200	60	140

SI units

Base and dimensionless SI units

Physical quantity	Name of SI unit	Symbol of SI unit
length	meter	m
mass	kilogram	kg
time	second	s
electric current	ampere	A
thermodynamic temperature	kelvin	K
luminous intensity	candela	cd
amount of substance	mole	mol
*plane angle	radian	rad
*solid angle	steradian	sr
*dimensionless units		

Derived SI units with special names

Physical quantity	Name of SI unit	Symbol of SI unit
frequency	hertz	Hz
energy	joule	J
force	newton	N
power	watt	W
pressure	pascal	Pa
electric charge	coulomb	C
electric potential difference	volt	V
electric resistance	ohm	Ω
electric conductance	siemens	S
electric capacitance	farad	F
magnetic flux	weber	Wb
inductance	henry	H
magnetic flux density	tesla	T
luminous flux	lumen	lm
illuminance (illumination)	lux	lx
absorbed dose	gray	Gy
activity	becquerel	Bq
dose equivalent	sievert	Sv

Decimal multiples and submultiples used with SI units

Submultiple	Prefix	Symbol	Multiple	Prefix	Symbol
10^{-1}	deci-	d	10^{1}	deca-	da
10^{-2}	centi-	c	10^{2}	hecto-	h
10^{-3}	milli-	m	10^{3}	kilo-	k
10^{-6}	micro-	μ	10^{6}	mega-	M
10^{-9}	nano-	n	10^{9}	giga-	G
10^{-12}	pico-	p	10^{12}	tera	T
10^{-15}	femto-	f	10^{15}	peta-	P
10^{-18}	atto-	a	10^{18}	exa-	E
10^{-21}	zepto-	z	10^{21}	zetta-	Z
10^{-24}	yocto-	y	10^{24}	yotta-	Y

Electrical and electronic circuit symbols

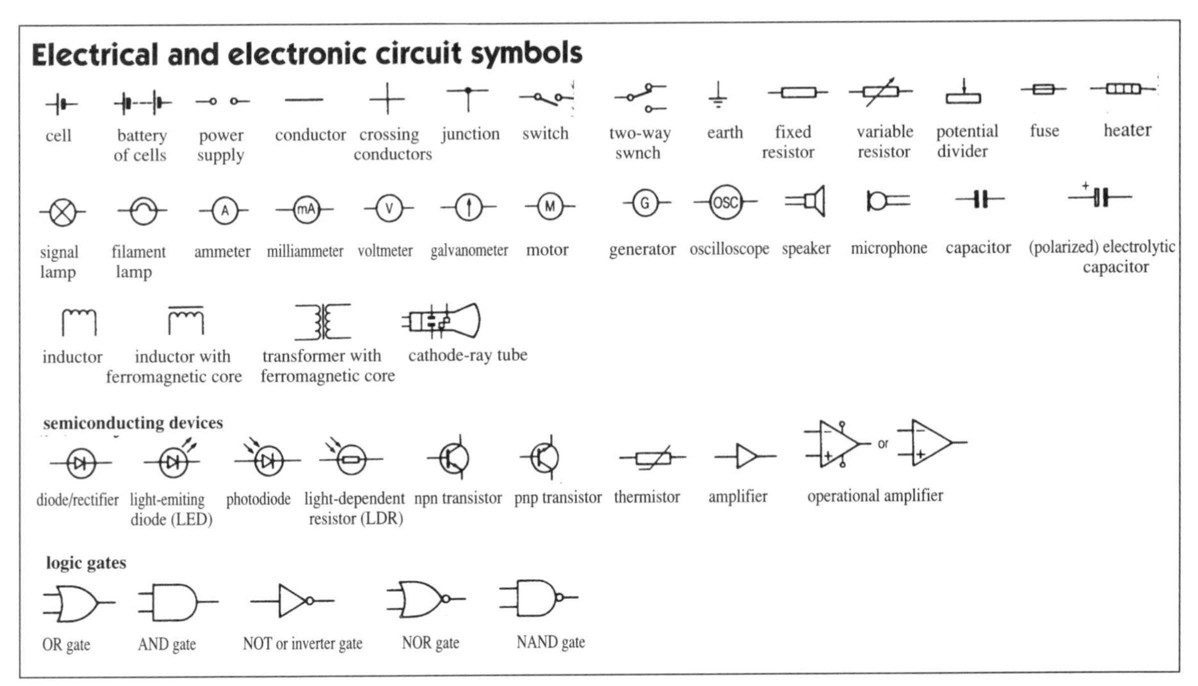

Electromagnetic spectrum

1 Some sources of electromagnetic waves
2 Spectrum of electromagnetic waves
3 A detail of the spectrum
A = ultraviolet
B = visible (violet, indigo, blue, green, yellow, orange, red)
C = infrared

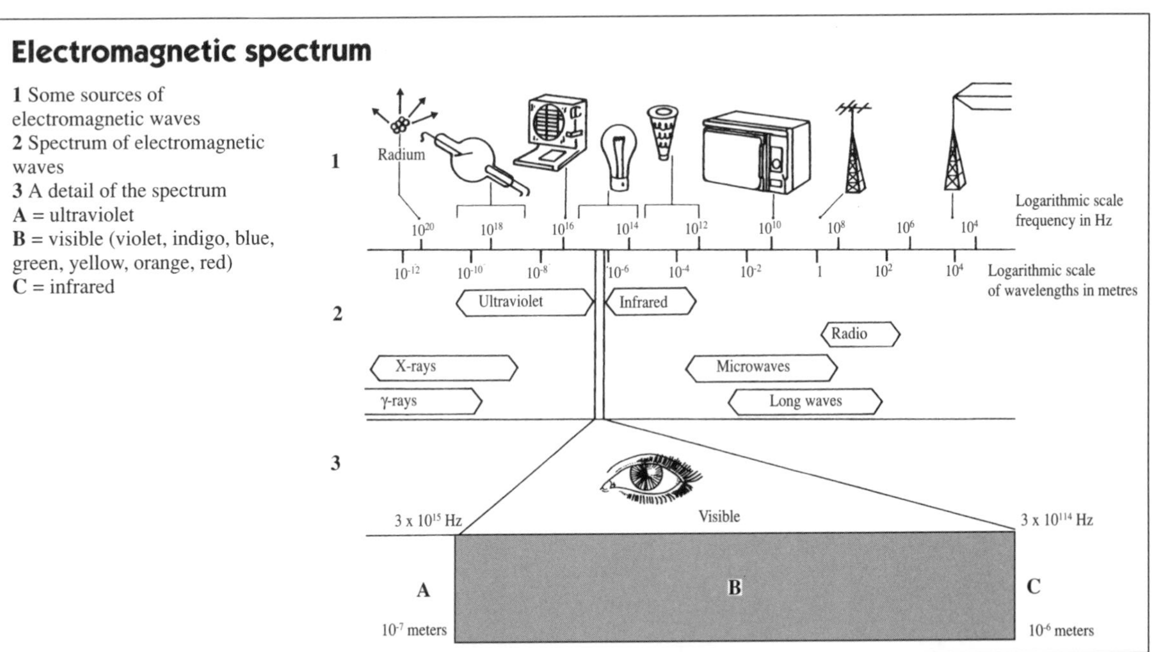

Types of energy

Relationships between forms of energy:

Type of energy	**Example**
1 Mechanical energy: kinetic or potential energy	**a** dynamo **b** turbine
2 Electrical energy: the energy produced by electrons moving from atom to atom	**c** motor **d** electric heater **e** light bulb
3 Nuclear energy: the energy stored in the nucleus of an atom	**f** power station **g** bomb
4 Heat energy: the energy produced by the random movement of a substance's atoms. The faster they move the hotter they become.	**h** hot air balloon **i** anything red or white hot
5 Radiant energy: consists of rays, waves, or particles, especially forms of electromagnetic radiation such as infra red radiation, light, ultraviolet radiation, X rays, gamma rays, and cosmic rays	**j** solar heating panels **k** solar cells **l** photosynthesis
6 Chemical energy: the energy stored in an atom or molecule and released by a chemical reaction	**m** batteries **n** gas or oil furnace

Lenses

Lenses and other transparent substances change the direction of light by bending (refracting).

Types

1 Double convex: used in magnifying glasses
2 Plano-convex: used in some slide projectors
3 Concavo-convex: used to combat farsightedness
4 Double concave: used to produce reduced images
5 Plano-concave: used with other lenses in cameras
6 Convexo-concave: used to combat nearsightedness

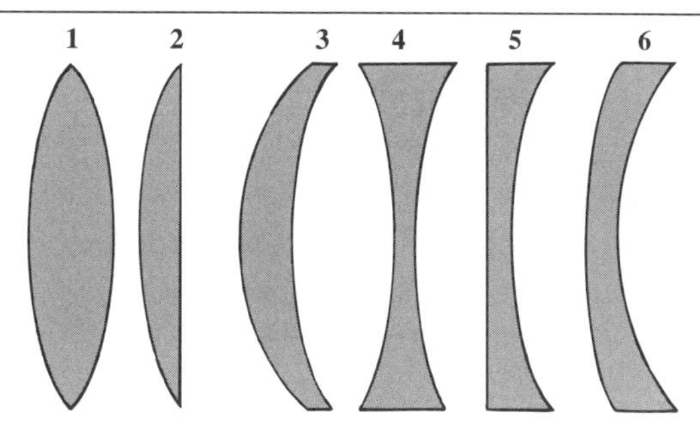

Concave lens

Thinner at the center than at the edges. It bends parallel light rays outward as they pass through the lens. It produces a reduced image, right way up, and appears on the same side as the original object (**1**). This is a virtual (not real) image and it cannot be focused onto a screen.

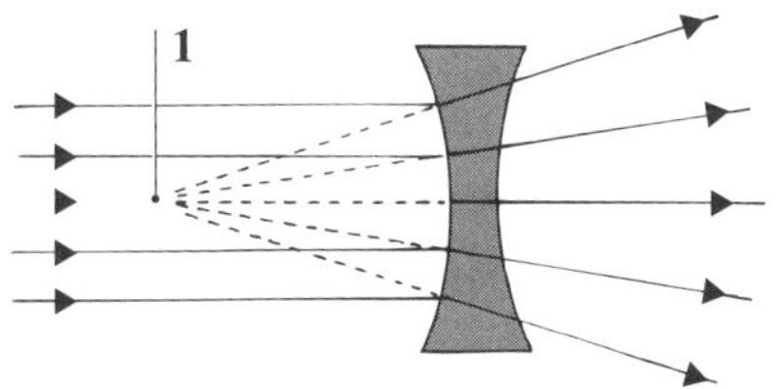

Convex lens

Thicker at the center than at the edges. It bends light rays inward to meet at a focal point (**1**), producing a real image. The distance from the center of the lens (**2**) to the focal point is the focal length. An object less than one focal length from the lens becomes an enlarged image on the same side of the lens.

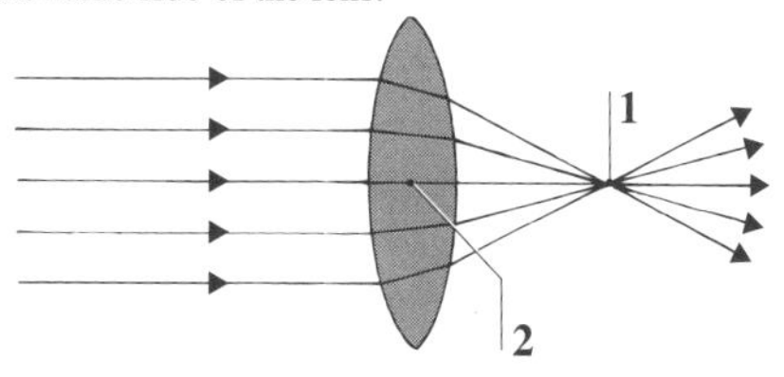

Radioactivity

Types of radiation

Name	Relative Mass	Charge	Nature
Alpha α	8000	+2e	Helium nucleus
Beta β	1	-1e	Electrons
Gamma γ	0	0	High-frequency electromagnetic radiation

Range in air

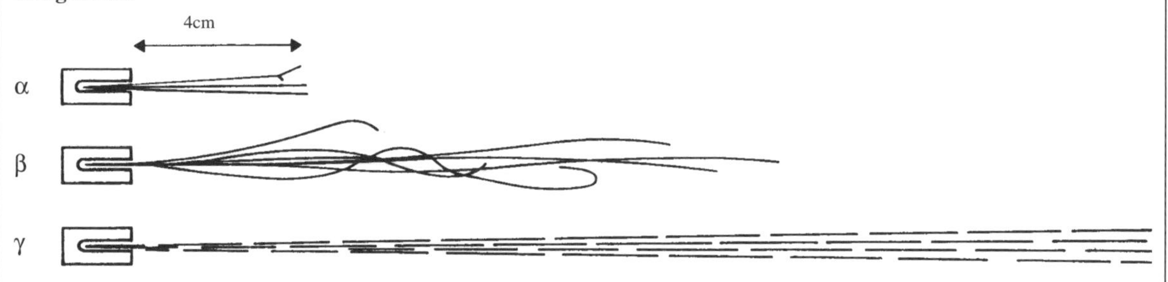

Penetrating power

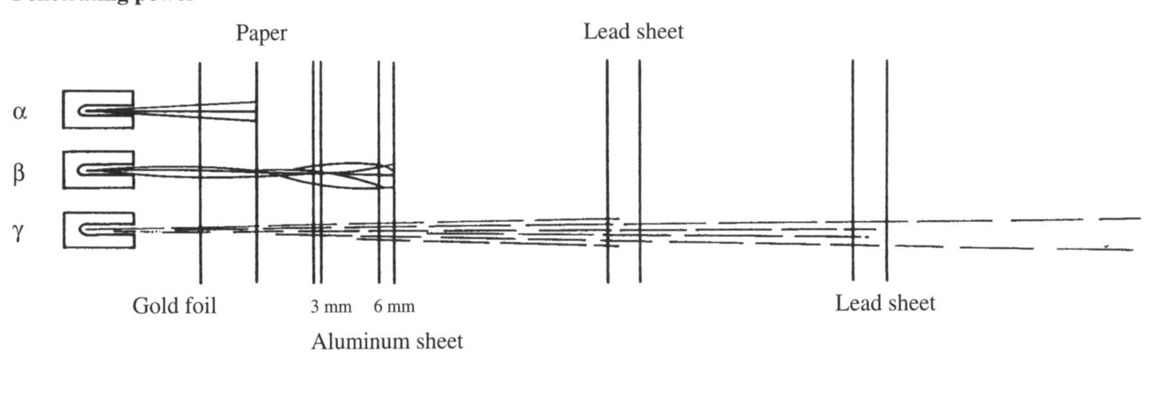